A PATH TO
PERSONAL FREEDOM

TIM CROSS

Austin, Texas

www.thearchitectureoffreedom.com

TABLE OF CONTENTS

Table of Contents

Table of Contents

DEDICATION

This series of books evolved from a simple letter to my daughters. If, early on, I had any real sense that this project would become a dominant part of my life for many years, it is quite possible that I would never have started. As with all heartfelt journeys, it unfolded moment by moment.

I therefore dedicate this effort to my two daughters, Emily and Rachel, and to Merlyn, their mother, who was my partner throughout much of the early part of this journey.

These books would not have been possible without the endless help and full, loving support of Connie Colten, my longtime best friend, life partner, wife and, most recently, editor, so, of course, I also dedicate this effort to her.

Many others have directly contributed to making this work possible. Thanks to all of you who have helped and who continue to be a part of my amazing personal journey. Ours is a continuous journey of discovering our connections, so I collectively thank all of us, together. We are so beautiful.

"Live life as if everything is rigged in your favor."
Rumi

"Whether you think you can, or think you can't—you are right."
Henry Ford

"We must be willing to let go of the life we planned so as to have the life that is waiting for us."
Joseph Campbell

"When I let go of what I am, I become what I might be."
Lao Tzu 600 BC

"We don't see things as they are. We see them as we are."
Anais Nin

"Life is what happens to you when you're busy making other plans."
John Lennon

"Out beyond ideas of wrongdoing and rightdoing, there is a field. I will meet you there."
Rumi

"A human being is a part of the whole, called by us 'Universe,' a part limited in time and space. He experiences himself, his thoughts and feelings as something separated from the rest—a kind of optical delusion of his consciousness. This delusion is a kind of prison for us, restricting us to our personal desires and to affection for a few persons nearest to us. Our task must be to free ourselves from this prison by widening our circle of compassion to embrace all living creatures and the whole of nature in its beauty. Nobody is able to achieve this completely, but the striving for such achievement is in itself a part of the liberation and a foundation for inner security."
Albert Einstein

INTRODUCTION

This book is about our universe, our most recent science, our spiritual search, our experience, and how all of these support the once crazy idea that we are here to live lives guided by the uninhibited expression of personal freedom. I define freedom as letting go of those things in our lives that block us from experiencing our birthright, which is living fully within the flow of this infinite creation—a flow that is shaped by a deep type of connectedness that we sometimes call Love. Our primary purpose is to explore, express, and share our individual and unique gifts so that all of us, together, can know and fully experience life's amazing possibilities. When we resist this calling our lives appear difficult, dangerous, and limited but, when we embrace our true purpose, everything changes and our lives begin to flow perfectly.

When I published my first book on this topic, *The Architecture of Freedom,* some readers requested a shorter version that would be easier to read and carry. This new book started as a response to that request but, for me, it has become so much more. The process of distilling these ideas has led to a powerful deepening of my own understanding, which I believe I have communicated in this volume. Because of the nature of abridgement, I will not include every topic from the original book and most of the topics chosen are discussed with less detail. My intention is to present just enough information to make this volume self-contained even though it has been distilled to reach a broader audience. If a reader desires more detail or discussion, they can refer directly to *The Architecture of Freedom.*

While the ideas in this book are not critical for living, they certainly can help make our lives much more enjoyable. We are here to explore every nook and cranny of our physical existence, and even if we change nothing, collectively, we will still be fulfilling this role perfectly. However, there are an infinite number of ways to experience life and, as this book illuminates, we each have the absolute power to choose our own personal experience. ***In fact, each of us is the only person that can determine our experience because no one else is even involved.***

2

Living our lives with a different awareness, one in which we fully understand that each of us is an important and integrated part of something much greater than our old and obsolete vision of separate competing individuals, will powerfully influence how our lives appear and feel. We literally have the ability to change the appearance of our entire world by learning how to change our deepest hidden beliefs.

This book was not written to convince skeptics. Both versions have been written as personal affirmations: verbalizations of a deeper type of knowing. The process of writing these books helped me to organize and more deeply understand my own thoughts; as I found more descriptive words for my experiences, my own understanding moved closer to the surface and became more focused. If a reader finds that my experience is familiar, then these books will certainly provide additional reinforcement for their own growing awareness; and the books might also contribute a few new insights and/or helpful guidance. These books explore universal and important ideas that are not often freely discussed in our contemporary culture.

For some readers, these ideas may seem like little more than science fiction, because much of what I write about is technically un-provable. However, being un-provable using the technical and conceptual tools we have at our disposal today, does not make the concepts any less real. ***Rational human thought, which relies upon words, concepts and thinking, may not ever be able to "prove" these ideas because, as we will discuss, our lives actually inhabit a much broader space that stretches far outside the limits of both our three-dimensional physical world and our rational thought processes.***

Ironically, while the scope of creation may lie beyond our understanding, it is not beyond our "knowing," and this is why inspired "art" of all types is so powerful—it somehow communicates this "knowing." No person can describe the depths of creation in a way that will help another understand it unless the other already "knows," at least at some level. Only then can our written words provide meaningful support.

Using words to describe, "that which is beyond words" is tricky business, so I employ a few tricks to help readers move beyond our traditional thinking. If I am assigning a slightly different meaning to a familiar word or term, or if I want the reader to reassess the meaning of some common term, I will use "quotation marks." If the term is technical, scientific, spiritual, religious, or philosophical, I will point this out by using *italics,* hoping to encourage the reader to look up the definition. If my use is atypical or unusual, I will define the first use of this word or phrase. At times, I have capitalized a term that is not usually capitalized to imply a unique meaning. For example, there is a difference between truth and Truth; truth is what appears to us to be true in any moment, while Truth is what exists at the very source of creation, far beyond the current limits of human comprehension.

Our perception of truth is always relative and changing; what is true for one of us on one day and in one place might not be as true for someone else, or even for us in another time or place. From within the realm of our existence—our three-dimensional physicality—it is impossible to proclaim an "absolute" truth, even if such a thing exists. I, therefore, make no claim that any of the ideas in this book describe an absolute truth. However, I am certain that the ideas described in this book are much closer to a fundamental truth than the current cultural paradigm that has been shaping our lives for the past 500 years. *At the very least, this book points to the general direction of a deeper relative truth.*

Human culture has evolved three distinct pathways for collecting, assembling, and saving knowledge. First, we gain knowledge from our direct, everyday experience: our own and the experience of others with whom we closely interact. This level includes our direct intuitive and spiritual experiences. Next, we have our cultural and spiritual traditions that have been passed down through the generations, including our religions. Lastly, for over 2,000 years we have built a great body of knowledge through formal scientific study, research and experiment.

While all of these three directions can help us to better understand our lives and creation, each direction, in isolation, also has aspects or parts that can seem incomplete or even wrong, sometimes leaving us unsatisfied or even disturbed. Historically, these three important information sources have often appeared very different and separate, at times even in conflict or contradiction with each other. *It is only recently that we have started to recognize that there is a substantial convergence or overlap: a common region where experience, spirituality and science actually share the same expansive vision.*

Since our spirituality, experience, and scientific study all represent our attempts to understand truth at a deeper level, the region where they intersect and overlap should provide the most fertile hunting ground for discovering this truth. In this and my original book, I explore this important common ground that is shared by these three great human endeavors. To this end I will discuss personal experience, my own and others'; our spiritual traditions, which are built upon cultural experience; and our paradigm-changing science of the last one hundred years. I will show how all three of these traditions point in the same direction and towards the same ideas– actually a single, unified idea that is disguised as many to work with our mode of thinking. While some of these ideas are technically un-provable because of their very nature, I will show them to be fully supported and reinforced by each and all of these three very different traditions. When we examine our cultural history in more detail, we find that many existing and ancient cultures have traditions that include the recognition of a similar expanded vision for life, traditions that closely parallel the discoveries of our most recent science.

In my attempts to describe what creation is, I discovered that it was much easier to understand and explain what it isn't. This is because of the limited and limiting nature of our conceptual thought and language. Describing "what it is" requires that we move beyond our logical thought process, and we will always lack effective tools for doing that. When discussing "what it isn't," the subject and ideas can all be contained using logical processes and words, and this makes it fully understandable and describable. Because of this, both books include extensive discussions of "what creation isn't."

The Architecture of Freedom, included descriptions of my own personal process, but much of that material has been removed from this book. While living my life searching for specific answers, I eventually discovered that it was the experience, and not the answers, that I was actually searching for. I learned that we are not here for answers or "to do it right." Instead, I discovered that we are here for the experience—not just for our own individual experience, but for creation itself, in order to grow and expand through our collective experience. This also means that while we are eight billion individual humans, we are also all part of one greater living "being." Like cells of a body, we all contribute to the whole: each and every individual experiencer is unique and a necessary part of creation.

The special nature of our dualistic, three-dimensional world is such that there must be *contrast.* Relationships built upon differences allow us to understand and move through our environment. Without hot, there can be no cold; without up, there is no down. Our universe is built upon this type of dualistic *contrast*; human life requires it. Our world can never be only good **or** bad, for good **and** bad must exist together; like all dualistic extremes, each completely depends on the other. In fact, our physical universe could not even exist without such extremes since every aspect of our physical universe is formed by the interaction of opposing forces. Because our existence actually requires both polarities and everything in between, the day and the night and the Ghandis and the Hitlers will always exist with each other. Ghandi, Hitler and many others, equally extreme, have only appeared in our world to fulfill these necessary roles for the benefit of all of creation!

After more than sixty-five trips around our Sun, I now find myself comfortable with this amazing process that we call life; the constant change and challenges of our everyday lives finally make perfect sense. *Today I reside in an easy, deep and durable peace, knowing that our world is not broken; in fact, it is far from broken—it is actually perfect in every way. I now understand that nothing outside of me needs to change, knowing that real change and movement can only occur from within.* The world exists as it does because it is the perfect vehicle for personal inner growth and human evolution. *The deep, holographic nature of reality, "as within, without, and as above, below," means that the physical manifestation of the outer world will always be a direct and perfect reflection of our inner world.* This means that any desired changes must begin within. Furthermore, I also now understand that personal change is not even necessary—it is just one more available option for exploration. All things in creation are perfect just as they appear to us, in each and every moment.

Of course, I have not always been this comfortable with life's processes; this level of personal awareness arrived gradually. The younger version of myself was very disturbed and troubled by what I saw unfolding in the "outside" world, and I often found myself blaming my difficult experiences on the "wrong" behavior of "others." I grew up in the American "picture-perfect" world of the 1950s, came of age in the 1960s, and entered adulthood while Nixon was president. These were extraordinary times for the shaping of lives dedicated to personal change and evolution.

I studied physics in college because I enjoyed and excelled in the subject at the high school level. For me, this more traditional physics was intuitive and I found it easy to understand. However, once in college, the study of *relativity* and *quantum physics* presented me with an unexpected challenge because these newer theories explore a territory that exists beyond our human ability to visualize or understand. I began to realize that the world, and therefore life itself, must be very different from our collectively imagined and culturally inherited vision. My focus then shifted from the actual physics to what this new physics was actually saying about the deeper nature of life and our universe. From that point on, to my mind, little else seemed to really matter. Over time, I realized that if I examined our spiritual traditions and my personal life experiences more deeply, they both also reinforced this same strange world view that I first discovered through my study of *quantum physics* and *relativity*.

Due to the limited nature of our conceptual thought process and the infinite and hidden nature of creation, we simply are not equipped to be able to understand the full depths of creation. A large portion of this book is devoted to explaining just why this is so. Even though we cannot understand these invisible and infinite extents, we can use their existence to help us imagine different ways to understand our lives. A more philosophical exploration of this topic can be found in *The Architecture of Freedom.*

As we explore this architecture, some readers will realize that many of these ideas defy "common sense," and they will be absolutely correct in their reasoning. As Einstein expressed so clearly more than one hundred years ago, "common sense is the collection of prejudices acquired by age eighteen." It is precisely this, our "common sense," which must be rethought and relaxed at the doorway of this new reality, for it is our rational "common sense" that keeps us firmly anchored to the old paradigm. It was our "common sense" that once convinced mankind that the Earth was flat and the Sun circled the Earth. "Common sense" is a great tool for physical survival, but not for exploring the deep mysteries of the unknown.

The idea of proof is a powerful and critical tool in modern science, but all proofs necessarily build upon "known facts and techniques," which means they are confined by the limits of our conceptual knowledge and thought processes. Proofs, scientific and mathematic, which are refined and organized through our system of logic, form the foundation of our scientific "common sense." They are always built from the bricks and mortar of that which we already understand.

While a mathematical proof will serve to extend our vision (for example, exploring new ideas within Einstein's four-dimensional *spacetime*), we can arrive at a point, within the deepest of journeys into the unknown, where our need for proof must also be set down. Proofs, by their very nature, tie us firmly to our old concepts and illusions. ***The greatest insights are often available only when we can let go of that which we think we already know.***

This book's foundation is based upon real science but, because some of the book's most important ideas are, by their very nature, un-provable, it is not a scientific study. ***It is, instead, a book that explores the important but rarely examined territory where our best science, our deepest spirituality and our most profound common experiences intersect.*** This place of intersection is of special interest because it contains elements shared by all three of these human traditions. That this common ground even exists indicates the proximity of a deeper truth: one that connects and unites much of our knowledge and history. All three perspectives point us to a similar understanding about the interactive, interwoven, and infinite architecture of creation: an architecture that actually depends on individual and personal freedom. Our universal home turns out to be a fully interactive and interconnected Web of Infinite Possibilities that operates outside of the kind of time that we understand. Our universe is so much more than just a place that we occupy; it also completely describes and defines who and what we really are. What we discover in the most profound depths is that our universe and everyone in it are actually one and the same.

The following principles describe this amazing architecture: **"The Web of Infinite Possibilities."** We will learn that this architecture exists at a level of creation that cannot be expressed accurately through words or verbal concepts, but we must begin somewhere. I only introduce verbal descriptions at this early stage to provide the reader with a preliminary outline. This approach should make these new ideas easier to follow as they are further developed in each of the three sections: science, spirituality, and experience. Because these concepts are completely undeveloped at this early point in the book, some of them may seem random, strange or even ridiculous. The rest of the book is devoted to helping readers understand this architecture, and to describing why I believe that these twelve principles are the best current description for how life and our universe work. If these ideas are unfamiliar, refer to these principles as you encounter new concepts; at some later point these principles and recommended lifestyle changes will begin to make more sense. *On the other hand, if you already understand these principles, then please jump around and explore this book in any order—you can approach it either way.*

THE "WEB OF INFINITE POSSIBILITIES"

The core concept of this book, the architecture of our universe, I now introduce through twelve fully interconnected verbal ideas, which only seem to be separate. I will continue to dissect these ideas and explain them using different approaches and methods throughout this book. All twelve represent different, three-dimensional interpretations (or views) of a single, multi-dimensional principle—they only appear to be different because of the way our minds need space, time and separation to function.

At this point in the book, these ideas may seem very strange or confusing. I only introduce them here as a reference point, so they can be explained and developed throughout the rest of the book. By the time *A Path to Personal Freedom* is finished, this vision of our universe should make perfect sense.

TWELVE PRINCIPLES

1. WE ARE INFINITE BEINGS: Thinking that we are only limited to these bodies is mankind's greatest and most confusing illusion. We are actually unlimited, eternal and infinite.

2. WE ARE MULTI-DIMENSIONAL: There exists, now and always, an infinite and fully interconnected structure, not unlike a woven web, built from the information for every possible outcome of every situation, choice, or thing that has ever been, or will be, possible. It is within this infinite fabric of creation that we exist, act and play—a perfectly designed structure that contains every possibility for everything that did happen or could have happened, ever. It exists, along with each of us, now and forever—everything actually exists "outside" of time. This Web is built upon more physical dimensions than our familiar three directional coordinates, but we are not able to directly "see" or experience most of this interwoven Web. It is invisible to us because it lies beyond the limits of our physical senses and our limited ability to conceptualize: it is actually not physical. This mostly invisible "Web of Infinite Possibilities" is the multi-dimensional and infinite fabric that forms the very structure of creation: the place where we live and express ourselves in all of life's various forms.

3. WE ARE VIBRATIONALLY BASED: Physical form is of secondary importance within this Web. Everything in creation begins with information that is communicated through vibration. The entire Web is alive with vibration. Just where in this amazing, vibrating Web our individual awareness expresses itself in any given moment depends on the resonant qualities of our individual soul. Physical form, as we recognize it, only appears as a secondary result of the way information is expressed and shared through vibration.

4. FORM IS AN ILLUSION: All of the forms that we encounter in our physical realm are only shadows or dreamscapes, even though they feel and seem very real. Our universe is created from at least eleven dimensions, but we can only participate in that smaller part of creation that can be "seen" from our own limited, three-dimensional perspective. Form is only created through the separation or division that unfolds within our three-dimensional duality. Our real and solid-seeming world is not what it seems and is only one small aspect of creation—beyond and within it, there is so much more. Because our bodies are form, they also are not who we really are: we are so much more. This is the illusion that creates the greatest amount of confusion for everyone.

5. TIME IS NOT LINEAR OR ABSOLUTE: Time, as we know and understand it, does not really exist. Our sense of time passing exists only because our three-dimensional minds must process information in discrete chunks and in a linear fashion: time is created to keep our particular type of "thinking" organized. All moments actually happen "outside of time" in the "now," but our brains can't process events this way. This "now" is accessible by us, and learning how to access it facilitates our conscious connection to that part of our being that exists beyond thought and time. The instant we think about anything or try to explain what it means, the "now" is lost—it instantly becomes the past or the future.

6. ALL POSSIBILITIES ALREADY EXIST: All of the potential possibilities for form and life are already expressed in an infinite number of other pre-existing "worlds." Some of these worlds are identical to ours; others are very similar; and most are very different from our "world." These "worlds" are positioned in this fully interactive and multi-dimensional Web so that all possible "worlds" are layered to be always directly adjacent and "parallel" to each other, allowing instant intercommunication and access. This is called enfoldment. From our three-dimensional perspective, we usually only have the ability to consciously observe one of these "worlds" at any moment of "time" but, as we learn to step "outside" of time, we will broaden this awareness.

7. OURS IS AN EVOLUTIONARY JOURNEY: We are an integral part of creation, designed to grow, experience and expand awareness in order to know and become one with everything in existence. Each evolutionary step is an expansion of consciousness as we gradually merge with what lies beyond. Each new level of our expanded beingness will always include all that came before. We never lose our sense of self; it just constantly expands to include more. This journey is the purpose of life.

8. WE HAVE THE FREEDOM TO CHANGE OUR EXPERIENCE: We can (and do) move continuously and automatically between these parallel "worlds." Every one of us is constantly shifting our viewpoint, or position, within this fully expansive Web. What we imagine to be our single "world" is actually many different parallel "worlds" that are continuously shifting, changing and shuffling. That which we understand as the "I" is always expanding, contracting and migrating between these different "worlds." In any given moment the "I" always manifests in that particular part of the Web that most resonates with our "fundamental core vibration," also called our soul. We each have the ability through personal transformation to consciously adjust, or tune, our fundamental core vibration; when we do, we resonate and appear within new locations or other "worlds" within this Web. In this way, we are always adjusting our experience by moving throughout an infinite collection of universes. Since these are truly different "worlds," these shifts can and do completely change the way our outer world appears to us. Because of the way our brains function by employing cognitive dissonance, smoothing and interpolation, most of the time we don't even notice these shifts unless we are particularly open or they are extremely dramatic.

9. WE EACH ARE SMALLER COMPONENTS OF A SINGLE VIBRANT LIVING BEING: Everything in creation is alive with vibrational life and intimately connected with everything else. We are so closely connected with everything that, at the deepest levels, we can say that there is really only one thing in the entire universe. We all interact together as a single entity, whether or not this is our awareness; our functioning as separate beings is helpful to facilitate the creation of exciting new possibilities. As we open to more of this amazing expansiveness, that part of us that we experience as the "I" will expand to intermix and co-join with "others," but the core awareness of "I" will always remain, no matter how we evolve. That "I" part of our being does not vanish— instead it grows and expands; however, as it does, it will continue to feel much like the old familiar self. Through this evolutionary expansion, we lose nothing except our limiting ideas and concepts. As we evolve more deeply, we expand, grow and deepen our interconnection with all "others." Our direct resonant connection to source is often called "Love." Love is actually a place of being— a place that we can always choose to dwell within. Love is that place where we can best experience our deep interconnection with everything.

10. LIFE IS ALREADY PERFECT: Due to this timeless interconnectedness, everyone is always in perfect harmony with the universe, and nothing is ever wrong or out of place. Since it is dynamic, self-reflective and completely interactive, the universe is always in perfect balance and this balance can be found in everything. It does not matter how our lives appear; they are always, in each and every moment, the perfect expression of being.

11. DEATH DOES NOT EXIST: Since time is not linear and form is an illusion, our concepts about the birth and death of our physical body are also only illusions. Birth and death represent one mechanism for shifting positions and journeying through this amazing Web. Death, like time, is only a three-dimensional concept. Our reflexive and unfounded fear of death is one of the main issues that inhibit us from fully experiencing this amazing process called life.

12. WE ARE HERE FOR THE EXPERIENCE: We are here in our physical lives to encounter, embrace and know every corner in our special part of this amazing Web. Our mission is to explore and participate in its every nook and cranny, no matter how wonderful or terrible it may seem. One reason for the appearance of many separate individuals is simply to facilitate this total exploration of creation. Through this diaspora of broadening experience and the eventual embrace and incorporation of everything that we encounter, together we will know "all that is." We will know true freedom only when we no longer experience ourselves as separate physical individuals, living within a limited amount of time, who are also expecting to die one day and then disappear. These beliefs are all part of the shadow illusion of our dimensional realm. This process of "knowing and becoming" through our opening to all of creation is the ultimate purpose of our strange and awesome journey that we call life.

These twelve statements represent some of the most relevant aspects of this hidden architecture that can be expressed in words. *A Path to Personal Freedom* was designed to fully develop, build, reinforce and explain these ideas, so that readers can better trust, understand and use this new awareness. *The Architecture of Freedom* covers these and many related ideas in an extended form. The reader must keep in mind that the multi-dimensional nature of the universe will always mean that it cannot be fully or accurately described using our three-dimensional concepts and language. Words, alone, will never be adequate for this understanding; ultimately, a deeper experience is required.

To even begin comprehending this vision of the universe, we must first break it down and focus upon smaller, more accessible three-dimensional ideas, partial views or perspectives. Therefore, throughout this book I will discuss these partial insights in different ways. As the reader progresses, a more integrated understanding of this new paradigm should begin to emerge through this collection of smaller glimpses. Eventually, in much the same way that we piece together jigsaw puzzles, the pieces will begin to fit as a map of this new territory forms in each reader's mind. This is exactly how early explorers of our world constructed the first physical maps of the once strange places they were exploring.

A temporary suspension of our normal rational analysis may help some readers with their process of integrating and understanding these ideas. While our rational minds are great tools for living, they always anchor us to our old ways of thinking—our existing paradigm. *It is critical that each reader arrives at the point where he or she understands (or at least temporarily accepts) the premise that the universe is infinite and multi-dimensional. The section describing the physics should help some to reach this important threshold. The sections on spirituality and experience will then reinforce and supplement what this science describes. Once the reader makes*

this leap, everything else presented within this book should easily fall into place.

I certainly do not ask readers to blindly accept the ideas in this book. Since many of these ideas are un-provable, that would be asking for an act of faith. ***Instead, I am suggesting that readers adjust their lives to better align with these principles, and then just experience how their lives change.*** The material in this book supports and explains the recommended lifestyle changes and the "real life" potential of this new vision.

The multidimensional Web described is the structural foundation of our universe, but it is not physical in the usual way that human beings understand physical things. It contains an *infinite* number of possible expressions for life, all utilizing *multiple dimensions* and *enfolded* in a way that puts every part in direct communication with every other part; everything in creation is always intimately interconnected and always fully intercommunicating and interactive. It would be more accurate to refer to its structure as an invisible and hidden ***Web of Interconnected and Infinite Possibilities.***

I recognize that as our knowledge of the universe deepens, it is probable that we will discover that the cosmos includes even deeper and more fantastic levels than those that we are discovering and discussing today. Because our universe exists in a "space" that is built upon many more dimensions than those that we now understand and perceive, we will always run into one particular conceptual roadblock. ***We are being asked to imagine things and ideas in our minds: a tool that is designed to understand and function in an environment of only three dimensions, far fewer than the actual number that these concepts and ideas are actually built upon. Throughout this book, I refer to this conceptual limitation as "the dimensional problem."*** Our human perception is contained within and is limited to our small region of creation. This is the single reason our lives seem to be surrounded by great amounts of mystery. We simply do not have the senses or conceptual ability to perceive much more.

Human beings have evolved to function efficiently in our three-dimensional environment. Being three-dimensional specialists, we have only a limited ability to imagine concepts that exist beyond our known paradigm. We did not develop the capability to function in more than three dimensions because this type of expanded ability was not particularly useful for the survival of our species, and species survival is what drives genetic evolution. ***Through the greater wisdom of evolutionary design, the deeper shape and structure of our universe has been intentionally hidden from our view.***

While the secrets and truths revealed through a larger dimensional awareness are not critical for species survival, they still are very useful for understanding our condition, individual happiness, inner guidance, and the evolution of our individual being.

THE LIFESTYLE CHANGE RECOMMENDATIONS

The following is a list of the recommended and practical lifestyle changes that will dramatically change our experience of life. To make this shift, there is no need to understand the physics or even express an interest in science. Believing any of this book's conclusions—mine, or those of others presented here—is not a prerequisite for this experiment. If a reader sincerely seeks to change his or her life, then all that is really necessary is to try out these fully related lifestyle changes.

As an experiment, just assume, for a set period of time, that the proposed vision of a multidimensional, fully interconnected, infinite and eternal universe is an accurate and functional representation of its structure. If the universe is built the way that I describe, then this alone will naturally suggest a different way of walking through life.

For this trial period, live your life from this new perspective and just observe what happens. See if your life—and the world you encounter—seems to shift towards something that feels much more peaceful, joyful, real, and empowering: something that maybe feels more like "you." The most basic principle for this experiment is *"attitude changes everything."* This is a very powerful idea and it is also clear to most of us that we have control over our own attitude.

Deep change is a not a fast or easy process, so patience is necessary. The attitudinal changes that will make a real difference cannot be accomplished by simply "changing" thoughts; instead, we must honestly examine and modify our deeper beliefs, both conscious and subconscious. These existing belief systems are often hidden and therefore invisible, and are always powerfully rooted within our culture and our paradigm. Most of our core beliefs are rooted so deeply that, over time, they literally became integrated into our physical bodies. To be effective, this process must create change at these deepest levels, and this takes time, focus, trust and persistence. As our core beliefs shift, our *resonant vibrational* patterns will also shift. ***We will discover that life is all about vibration and that absolutely everything that we perceive begins and communicates with vibration.***

When we start to integrate these lifestyle changes, the first shifts will generally be very subtle. We may only sense a small shift in the ambiance or tonality around our lives—a delicate change of flavor. Noticing this initial shift could encourage a dive into deeper waters, and here our core belief systems can be touched. We must understand that living this way is a process: a lifelong process, actually an *infinite* living process.

As already stated, there is absolutely no need to understand the physics. There is also no need even to believe that the vision of the "universe" presented in this book is "real." The only thing necessary is to integrate these lifestyle changes into our lives and then these changes alone will eventually have a dramatic impact on our core *vibrational being*.

Some of these changes are ***to***:

- ***Live our lives without fear, especially our common fear of death.*** Be courageous and fearless in our adventure of self-exploration.

- ***Live our lives as if we all are a part of one being.*** Treat all others as we wish to be treated and harm no others intentionally.

- ***Know that Love is always available.*** Quit the search outside of ourselves for this Love. Love does not originate from the outside and it is never dependent on another person. Love is a place inside: a place that we can visit any time we are open and ready for the experience. Love is also a wonderful and special space to share with others. We miss the richness and depth of this experience only because of our learned choices and habits.

- *Live as if we are creating our universe anew with every deep thought and feeling.* We actually are. Speak and act with clarity and awareness. Remember that every thought or word is vibrational and has great communicative power; our words instantly communicate and connect with everything in the universe. We are never alone and everything we do or think has ripples that "stir" the cosmos.

- *Feel inside more deeply.* Learn to find and feel into all of our empty, dark and frightened spaces—individual and collective. This practice illuminates the darkness and helps our spirit to shine brightly.

- *Learn also to feel "others" deeply.* Deeper within, all others are only different expressions of the same self; others can then be experienced as the expansion of oneself. Develop empathy towards others as a deeper form of self-awareness.

- *Practice forgiving.* Forgive ourselves through learning to forgive all "others." Ultimately, there is no difference. Examine every hint of blame. At deeper levels, all judgment and blame are forms of self-loathing. Forgiveness and freedom are closely related.

- *Always remember that we are so much more than just our "body and mind."* Do not become lost in the illusion of the small self: the single body called "me." Meditate, breathe, and learn how to feel those parts of being that lie beyond.

- *Live as if there is no such thing as a secret.* Everything is so interconnected there is actually no place to hide a secret—all thoughts and actions are always expressed within creation. Learn to be impeccably honest with yourself and all "others." This is how the universe works; life flows much more easily when this truth is recognized.

- *Avoid waiting for, or expecting, others to change.* Remember that no other person needs to change for your world to change. In fact, you are the only person that can make this change—you are the only person who can ever change your world! The help you are looking for is within, not without.

- ***Live as if there is no such thing as a mistake.*** Instead, there are only new and different opportunities to deepen and enrich our experience and awareness. Every type of experience presents a chance for learning and an opportunity for growth.

- ***Remember that everything changes instantly with our participation.*** Our focus, intent, and attention completely influence the unfolding of events. Do not pre-judge a potential experience based on past experiences. Every experience is unique because our participation and attitude influence the unfolding of events.

- ***Practice finding the "positive" aspects that are inherent in every situation***. All things in life can be viewed from many different perspectives. Never hide from the "negative" aspects—they are equally important for balance in our *dualistic* world. Remember that our world of *duality* is built from opposites: there must always be a natural balance in life. Our physical world requires contrast, therefore all things have "positive" and "negative" aspects. Recognize, understand, honor, and embrace this necessary balance in all things.

- ***Stop rushing through life.*** Because "time" is not what we once thought it was, there is actually no reason to hurry. Slow down and savor each and every step of the journey.

- ***Always remember that everything is perfect.*** Life is perfect, exactly as it is, right now. Each and every moment is perfect.

- ***Give loving attention to whatever is right in front of us in every moment.*** Live fully in each and every "present moment." Life automatically presents each of us with our perfect opportunities for growth. Dive into each of these present-moment gifts from the universe.

- ***Get out of our own way; allow for a more natural flow.*** Stop sabotaging life with pre-conceived ideas. Once again, embracing what is right in front of us in every moment allows us to flow perfectly with the organic changes of life.

And...

- ***Always be grateful for this gift of physical life, no matter how it may appear in any given moment.*** Life is an amazing opportunity and gift. Develop an enduring *"attitude of gratitude."*

From these practices, alone, we begin to learn how to love "others" and ourselves in surprising new ways, while fully engaging in and enjoying the infinite process of life. *At first, this way of walking through the world is not easy; it creates many new challenges for our conditioned minds.* However, once this new approach is integrated and the natural inner shifts begin to happen, everything about our lives will start to change in wonderful and unexpected ways.

Commit to trying these lifestyle changes for a fixed period of time–try six months; there is really nothing to lose. If you try these and they don't work for you, the worst that might happen is that your world does not appear to change. There is no potential downside to this experiment.

These changes, put into daily practice, will initiate a deep and personal process of opening. This process will seem difficult at times– sometimes it will even seem impossible. However, at other times, life will flow so easily and naturally that we will find ourselves wondering why it ever seemed difficult. Growth is rarely an easy or straightforward process. As young children, we had "growing pains." We banged our heads and skinned our knees because we did not yet fully occupy our bodies. In a similar way, we are now discovering and learning to occupy more of our potential *being*. Any frustration and difficulty we experience are only "growing pains."

Most of this growing involves discovering how to drop our resistance to the natural flow of life. This also means that we become more available for whatever opportunities life might be presenting. This is an especially difficult change for the Western-trained mind because our culture has always taught us to be "in control" of outcomes. True availability includes openness to any and all possibilities, even those we would normally avoid. *It seems counterintuitive, but our lives are actually more "in control" when we are able to entirely let go of our habitual need to control outcomes.* Instead, we become more available to fully experience the amazing gift that is always unfolding before us.

Many writers, speakers, and workshop leaders have encouraged similar lifestyle changes through related approaches or technologies. Some of these people and methods are listed in the Resource section of the book. Try any of these techniques, for there is no perfect or right practice. They all evolved from the same or similar worldview, and they all point to the same place: a life steeped in a deeper truth, a life of freedom. The greatest truths lie outside the limits of our ability to sense, conceptualize, or even imagine, so if we desire to experience this deeper level of truth, we must be willing to explore beyond what we once understood as real.

WE ARE ALL REALLY ONLY ONE

Throughout this book, I use the term "individual," a word that implies separation. As we will discover, this separation is only another artifice of our language, culture, and limited dimensional awareness. I use this term with full knowledge that as our perspective evolves to include more breadth and a deeper understanding, everyone will come to understand how directly connected we all are. At our root, we each are only different-appearing aspects of a single vibrating being.

Our Western perspective encourages us to see ourselves as separate, competing individuals but, at deeper levels of truth, we are only as separate as the leaves of a tree. Up close these leaves might appear to be individual and in competition with each other for light and nutrients, but when our perspective expands to include the trunk and roots, we see these separate parts as integrated and important components of a single, grand organism.

Like leaves, we also are fully interconnected, but we don't usually see this connection because it exists beyond the veil of our three-dimensional worldview; it lies outside of our senses and ability to conceptualize. Our inter-connectedness also extends beyond humans to include all plants and animals, along with everything that exists in creation. As we evolve and expand our awareness, these various name- and form-based separations will dissolve as we gradually embrace our full, resonant place within creation.

We all think that we are separate and competing individuals, but just like the leaves on a tree, we share more than we can imagine.

LIFE IS FOR LIVING

As the reader attempts to piece together this vision, it is important to remember that it is not necessary to understand or believe these ideas. Our primary purpose in life is the adventure of "living." *We are living our lives to their maximum potential if we live fully in every moment—nothing else is necessary. Life is about engaging with what is presented in each moment.*

Life is designed perfectly for maximizing our engagement and insights, and this book explains why it must be this way. Our lives are always in perfect lock-step with the totality of creation because that is the inherent nature of this architecture—it is fully interactive and responsive. *Our lives may not always look perfect or complete, but they are exactly that*, in each and every moment.

A well-disciplined mind can be a wonderful aid for this process, by helping us focus on the "present moment" and guiding us back when we wander. However, our minds can also become our greatest obstacle when they bind us to our preconceptions or direct us away from those potentially difficult situations that our soul might actually require for its own deepening.

Only through this journey of living and expressing our individuality will we come to know the entirety of all possible human experiences. Through our collective participation, we witness and integrate all our human experiences—all of which are critical and of equal value to our *being*. ***Through this process called life, we share this experience, together, so that we collectively can come to know "all that is."*** When Jimi Hendrix sang "Are You Experienced," he was asking if we are open and available for this amazing adventure. This is our purpose. This is life's purpose.

SECTION ONE–THE SCIENCE

INTRODUCTION

With recent advances in our knowledge and technology, scientists are now able to peer further out into space, deeper into the composition of the tiny things that make up matter, and backwards and forward through time. What we are learning is astounding and life changing.

The minimal science in this book should not create fear; many non-technical readers of *The Architecture of Freedom* have described the science chapter as their first encounter with an understandable explanation of relativity and quantum physics. Since there is even less technical description within this book, even those readers who have avoided scientific explanations in the past should be able to understand this section. On the other hand, if a reader desires a more detailed understanding of the physics than either book allows, refer to the substantial resource section at the end of this book.

The last one hundred years of science have completely changed our understanding, so I will focus on these relatively recent breakthroughs. Some older scientific history is provided to add context. While science has been a trustworthy guide for my own process, understanding this science or its meaning is not necessary for our evolution—it will unfold regardless.

SCIENCE AND PARADIGM SHIFTS

The scientific discoveries derived from *quantum physics* and *relativity* have been rapidly and thoroughly changing our lives and culture for more than seventy-five years. Although conceptually different, this new physics has consistently proven itself to be extremely valid and reliable. It has also resulted in the production of a dizzying array of very practical, real-world devices. From this "common sense" defying physics, we have discovered, developed and refined transistors; laser; amplifiers; neon, fluorescent and LED lights; microwaves; CDs and DVDs; computers; atomic bombs and atomic energy. Using this revolutionary science, we have predicted new elements and particles, launched accurate space probes, developed accurate GPS systems, and achieved a much more detailed understanding of our "physical universe." Today, as much as one-third of our national Gross Domestic Product (GDP) is dependent upon or built from products based upon *quantum theory.* These once unimaginable, high-tech products that were developed from this theory have become as much a part of America as apple pie. *Quantum theory* is a critical and integrated part of our lives. It is as real as anything else is in our world!

Along with all of these practical devices, the last one hundred years of physics have also uncovered a large number of principles, interpretations and possibilities about our existence that simply do not make sense when viewed from within our existing, but now very aged, *paradigm.* Our world is still operating from an old worldview, based upon a scientific understanding that is now more than 500 years old and obsolete.

Today, we are actively witnessing the emergence of a new and life-changing global paradigm. The magnitude of this shift will be even more extreme than the changes introduced by the last great shift, the one that overturned our previous "flat earth" paradigm. We must remind ourselves that only five hundred years ago, before Newton, Galileo, Kepler, and Copernicus, the "flat earth" paradigm was still the dominant worldview. That long-dominant worldview did not disappear quickly or easily, as many deeply committed scientists, explorers, rulers, and religious leaders gave their full efforts, and often their lives, to help break this conceptual and cultural "flat earth" shackle.

There was then, as there is now, an established, organized resistance to dramatic societal change. Powerful and enduring resistance to change is natural within living systems. Since large cultural changes require enormous societal redirection or restructuring, they will be strongly resisted by the establishment and other power structures. This is an understandable response because corporations, institutions and individuals are deeply vested in the status quo and do not want to lose their fortunes or privileged positions. Built-in resistance is not a "bad thing" for society since it functions as a buffer, keeping society operational by reducing or *damping* radical fluctuations and instabilities. This resistance can and should be understood as a normal, organic part of the process of change. Functioning like the shock absorbers of a car, this built-in resistance can make the ride smoother for all of us.

Six hundred years ago, the "flat Earth" paradigm with the Sun obeying a god's daily command to rise on one side and set on the other was just a part of life, and it was impossible for most people to even imagine a different model. In 1543, Copernicus described a different solar system in which the Earth was no longer the center—physically, philosophically or energetically. With this shift, humanity instantly lost its self-proclaimed position as the most central and important element of the entire universe; we had to accept, instead, that we were a smaller part of a much greater whole. The public had no context from which to visualize this type or degree of change. Since this new worldview directly threatened the Christian church's absolute authority and hold on power, it was officially declared to be blasphemous. Many brilliant and courageous scientists were persecuted by the church for introducing or spreading these "dangerous" and heretical notions. The once-universal belief that the Earth existed as the center of the universe could not, and would not, die easily.

Galileo and many others tirelessly and courageously developed and presented additional clear, physical evidence for this new vision of our solar system. These cosmological pioneers and the new science that they brought to light continued to suffer terribly at the hands of the church and state, but eventually this new worldview began to be seen as potentially advantageous by some of those in power. Due to the possible economic benefits associated with the plundering of a much more expansive planet, explorations started to be encouraged by a few rulers. Because of the potential economic windfall, the church largely blessed these projects. Eventually, as real-world financial rewards flowed in, adventurers like Columbus and Magellan were granted state- and church-sanctioned permission and financial support. ***Thus began the full-scale exploration of a "new" territory that once only existed beyond the edges of their old paradigm.***

The first brave explorers proved that the world was round and not flat by physically sailing "over the edge," and then returning. For the public, this new paradigm became increasingly real as more and more riches and wondrous items flowed back to their old world.

During this time, Kepler, and then Newton, were also refining the physical laws of motion for planets and other objects. Their contributions did a remarkable job of describing why large physical objects, such as apples and planets, behave as they do; but, even so, the global integration of this new paradigm still took many generations. As the world adjusted to this new paradigm, humanity's understanding of the universe—and life itself—was forever changed.

Today, this now 600-year-old idea is still our dominant cultural paradigm. Once again, we are facing the same type of conceptual challenge: our science, real-world experiments and economics are pushing our old paradigm to the point of dissolution. Again, we find ourselves in the middle of a confusing process; driven by science and economics, our culture is dramatically rethinking its now-obsolete worldview. *We are just now beginning to learn how live with our new paradigm—The Architecture of Freedom.*

THE VISION SHAPED BY SCIENCE

The following is my list of some of the most significant ideas that can be deduced from the last one hundred years of science. After this list is a brief discussion of the science that supports these statements. A more thorough discussion and analysis can be found in *The Architecture of Freedom* and many other books, some of which are listed in the Resource section.

1. *Everything that we perceive and experience begins with an energetic stimulus to our sensory systems, which then must be processed by our brain and nervous system.* We know our human biological systems have their limits: for example, we cannot sense radio waves or hear, see, taste, feel, and smell things that dogs and other animals can sense. There is a lot going on just within our three-dimensional world that we miss because we lack the receptors and neurological systems to process certain signals.

2. *Time does not exist as we commonly perceive and understand it.* The "linear, one-directional time" that we perceive is only nature's method of organizing and limiting information so that it can be effectively processed by our brains. Everything already exists, not *in* any time but *outside* of time.

3. *Our familiar physical world of three dimensions is only a very small piece of a much greater expanse of*

creation. We know from Einstein's work that there are at least four integrated dimensions. For more than one hundred years we have known about the existence of four-dimensional *spacetime,* but humans still perceive only three dimensions (plus a layer of one-directional time). Some of our most recent science indicates that it is possible, even likely, that we inhabit a universe built upon at least eleven dimensions. Our physical experience is restricted to a very limited portion of creation: our three-dimensional realm.

4. *Much of life's mystery is the result of artifacts from this deeper dimensional geometry: a geometry hidden from us.* For example, in three-dimensional space we perceive *gravity* as a *force*; but Einstein demonstrated that it is actually the indirect way that we experience the curvature of four-dimensional *spacetime.* Other experiences, such as our sense of passing time, similarly are misunderstood due to how we interact with this hidden geometry.

5. *We experience and understand only about one percent of the "known" physical universe.* The more we learn about the cosmos, the more mystery we uncover, and those parts that we know we do not understand only increase. We are trying to understand the "timeless infinite" using the constraints and language of our time-bound realm. This can make a very simple, multi-dimensional idea appear very complex and therefore mysterious. Mystery also exists because there are vast parts of the Multiverse that we cannot perceive.

6. *Every part of our universe is in constant and instantaneous communication with every other part,* even though there might be trillions of miles or billions of years separating these parts. Many physicists and philosophers believe that it is as if there is only one thing in the universe. In three-dimensional terms, it is as if "every part is always in direct contact with every other part." *Our universe is fully interactive and always behaves like a single organism.*

7. *Whenever we participate in an experience, we influence the outcome.* Before our involvement, an infinite number

of outcomes are possible; but once we engage, a single outcome becomes "solidified" and appears as "real." Our expectations influence the outcome that we eventually experience as real. Our world reacts to us and would probably not even exist or function without our conscious participation. In one sense we are creating our universe as we go—on the fly.

8. ***Einstein demonstrated that everything is energy in one form or another.*** We can't see, understand or directly detect about ninety-nine percent of the *energy* in our physical universe. Our internal sense of the world as solid and material is an illusion created by our brains and nervous systems. Our individual sense of physicality and separate selves is also part of this illusion. It is a convenient illusion because it supports and contains our physical life in this realm.

9. ***Vibration is the universal language of creation.*** It is likely that the most basic thing in creation is *information,* which is then shared and communicated through various forms of vibration.

10. ***Our universe is infinite.*** The latest cosmological research strongly supports earlier evidence that our universe is infinite. This means that somewhere in this infinite expanse there must be an infinite number of "parallel worlds" that are exactly like Earth, with humans exactly like each of us. The more we learn about our cosmos the more this extremely strange conclusion appears to be accurate.

FLATLAND: A NOVEL FROM 1884

The most important idea in this book is that the space we occupy includes more dimensions than our familiar three. Before jumping directly into the science that leads us to this idea, it will be helpful to learn a little about dimensions and just what they represent. Since art often predates science, I will borrow from a Victorian-era novel, *Flatland,* to introduce this important concept.

It is not possible for any of us to directly visualize the expanded space that exists beyond our known three dimensions—we do not have the senses, tools, or conceptual abilities to understand the vast extent of creation. What we do have, however, is a direct exposure to and a detailed understanding of one-, two-, and three-dimensional space. Using this knowledge, we can observe relationships that can help us understand more about the nature of space containing more than three dimensions. The transition from one-dimensional space to two-dimensional space and the transition from two-dimensional space to three-dimensional space are completely comprehensible. Once we analyze these transitions, we can *extrapolate* to gain a sense of what types of things and relationships we might expect from the next transition: the shift from three dimensions to four.

This process is similar to *reverse engineering,* where a product is taken apart by others to determine just how it was built. In a similar way, we can go "back" to two-dimensional space, and then look at how things change as we expand to three-dimensional space. By backtracking to a world of only two dimensions and then observing what is discovered when we reintroduce the third dimension, we can predict many possible aspects and qualities of the transition from three dimensions to four.

This is exactly what E. A. Abbott did 130 years ago in his 1884 novel, *Flatland—A Romance in Many Dimensions*. I find it particularly interesting that this novel was written when Einstein was a child. When I first read it as a nineteen-year-old, I experienced a dramatic shift in my awareness as I first became aware of and then became mesmerized by the "dimensional problem."

While this work of fiction predates Einstein's theory of *special relativity*, it does not predate the original math that inspired Einstein's physics. Fifty years before Einstein presented his Theory of Relativity, the foundational mathematics for his future theories were described by Bernhard Riemann, James Maxwell, Marcel Grossman and several others. **Flatland is just one example of extra-dimensional space being described well before Einstein's Theory of Relativity codified four-dimensional spacetime.**

Abbott specifically wrote *Flatland* to make two very different points. On the social front, it was designed as a clever and cutting criticism of the Victorian social caste system; but in the scientific realm it introduced the general public to the new idea of extra dimensions.

Abbott's book does an excellent job of helping readers understand the conceptual and social challenges of a society that is meeting a new spatial paradigm for the first time – exactly what we are doing today.

Abbott carefully takes the reader into his creatively constructed, two-dimensional, Victorian universe called Flatland. He fully develops a very believable world, while gradually helping the readers understand and feel the limitations of his characters' two-dimensional mindsets. He does this so well that some readers even start to think and feel like his spatially restricted, two-dimensional characters do. Some, like myself, might even have the experience of feeling "trapped" by the unexpected constraints of this reduced version of the universe.

Once this two-dimensional perspective has been fully internalized by readers, Abbott then has a three-dimensional object (a sphere) visit Flatland. The sphere's intention is to introduce the idea of three-dimensions to one of Flatland's citizens, the book's main character. Through this character's limited two-dimensional perspective, we make our first acquaintance with this extra-dimensional object, the sphere, and we also begin to see and understand the "dimensional problem." Later, we witness our two-dimensional hero's struggle to explain to other Flatlanders his encounter with this three-dimensional being. The author's setting and plot help us to more easily understand just how difficult, or even impossible, it would be to describe three-dimensional objects from a two-dimensional perspective.

Through this cross-dimensional interaction, we easily see that, from the perspective of someone living a two-dimensional existence, three-dimensional objects, even simple ones, will always appear as very complex and inexplicable phenomena. At the end of the novel, the reader is then asked to imagine how difficult it would be for three-dimensional humans to understand and describe a world of more than three dimensions.

A Simple Experiment

The inhabitants of Flatland exist on a tabletop-like surface where every person and object is completely flattened. Theirs is a two-dimensional world because there are only two directions (*coordinates*) for movement. Inhabitants can move forward or backwards, left or right, or in some combination of these; a Flatlander cannot look or travel up or down because these directions do not exist. *In fact, they cannot think of or even conceive of an up or a down. From the Flatlander's perspective, this third dimension does not exist because it lies beyond their senses and outside their ability to comprehend: it is completely invisible.* Everything within Flatland is understood from the level of the tabletop, so a Flatlander sees only the objects or parts of objects that are in direct contact with the table. Any object that is above or below the surface of the tabletop will not be visible or discernible. If a coffee cup is sitting on the table, all that can be seen is the ring of the cup bottom that actually touches the table. *From the Flatlander's perspective, all parts of the coffee cup above the tabletop, including the coffee, do not even exist, because they lie outside of their conceptual horizon.*

To get a feel for their two-dimensional worldview, sit at your kitchen table, place a chopstick or toothpick on the table and then lower your eye to the table edge until you are looking at the object from the height of a tiny insect walking on the table—you might even imagine that you are a very small ant. Close one eye and then slowly spin the stick and observe. What you see is a line that gets longer, and then shorter as it turns or rotates. Next, cut a circle, a triangle and a square from a piece of cardboard, all about the same size, and place them flat on the table. Close one eye and lower your open eye so you can see only the thin edges of these objects, and observe. The circle, triangle and a line should all look about the same because we only can see the closest edges of each object, and we have no depth perception because we are using only one eye. (While this loss of depth perception is not a requirement for a two-dimensional worldview, it does add another useful experimental condition; by sacrificing our depth perception, we can better understand what it means to have sensory limitations.) This demonstration illustrates how critical some of our abilities and senses are for perceiving and understanding our own worldview. *If we lack dimensional perception or certain senses, we miss many things, even if they are right in front of us.*

Life in Flatland

In Abbott's novel, the Flatlanders lack depth perception, so from their perspective circles and triangles all look the same. As we observed in our tabletop experiment, unless they rotate, all shapes look the same, like a line. Flatlanders recognized shapes by touching or feeling the sharpness of their points and would describe different shapes as "feeling different." As we discovered in our experiment, when triangles (i.e., the common working class people of Flatland) turn, they become shorter, then longer, then shorter again, and so on. Rotating squares look much the same as triangles, except their changes in length would be more frequent for the same number of turns. When circles (i.e., the special priest class of Flatland) turn, they don't change length; they always look like same.

Flatlanders describing the difference between triangles and squares, therefore, could also introduce the idea of "time." For example, Flatlanders might say that a spinning square changes its size more frequently than a triangle. From our broader, three-dimensional perspective, we can easily see that it is only the geometry that is different; "time" has nothing to do with their differences. "Time" is introduced only because of the perceptual limitations of their two-dimensional geometry. Since Flatlander's senses and mindset have no understanding of "above," they cannot observe these simple shapes from above and count the sides as we can. To do so would require the use of an extra dimension—the third dimension. This is just one example of *dimensional limitations* that lead to the "dimensional problem."

In Flatland, residents describe the differences between shapes by talking about **feel** and about how things change with **time**. Time and feelings are two complicated ways that Flatlanders describe some quality which, for us, is a simple a difference in shape. If they could just add one more dimension to their perspective, they would easily see that these different classes of citizens are all distinct shapes: triangles, squares, rectangles and circles. They would then realize that it is not necessary to use **time** or **feeling,** to describe these objects.

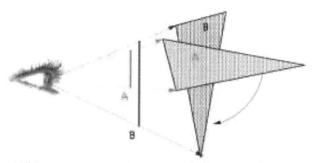

In Flatland the inhabitants lack depth perception and only see what is in their flat plane. Looking from above it is easy to see that triangle A and triangle B are the same shape and size. Their only difference is one is rotated. Flatlanders can not see from above. All they perceive is the line A and the line B, which are different lengths. As the triangle turns the Flatlander perceives a line that changes length as "time" passes.

A Sphere Visits Flatland

As previously mentioned, once Abbott fully establishes the nature of ordinary life in Flatland, it is then visited by a three-dimensional sphere. Imagine yourself living in such a flat land as the sphere comes to visit you by moving through the unseen three-dimensional space, before arriving at your tabletop world. When it first makes contact, that is when the sphere just sits on the table, only the bottom tip of the sphere actually touches the surface of your tabletop world. At this first moment of contact, observers in this flat world would only see a dot—remember that the Flatlanders cannot look up and see the rest of the ball that is sitting above the tabletop. All that they can see from their perspective is that single point where the sphere touches the tabletop (set a ball on a table and take a look).

You must imagine this sphere as a magic ball, one that can pass right through a table, or as a ball that is slowly being dipped into a flat pond. (This second image might be easier for many readers to visualize.) The sphere continues to move through Flatland, but as the sphere passes through the plane of the tabletop (or the top surface of the water), that initial single point of contact grows to become a circle. As the sphere or ball continues to move, the Flatlander will see the circle growing bigger because a thicker part of the ball is now passing through the tabletop. To the Flatlander, who lacks depth perception, this circle looks just like a line that gradually grows longer over "time." Once the equator or midpoint of the ball passes through the tabletop, the circle will then begin to get smaller. At this point, from the Flatlander's perspective, the line starts getting shorter, until, once again, it becomes only a small point or dot; this is what they see at the very last moment of contact.

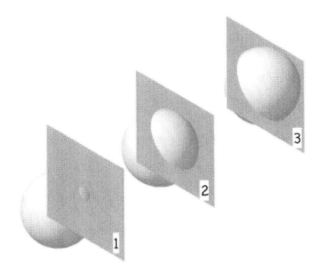

A three-dimensional sphere passes through the flat, two-dimensional world called Flatland. Flatlanders can't see above or below their flat plane of a world. First they see a dot which becomes a very small circle (1). As time passes the circle gets larger (2) until it reaches the middle (equator) of the sphere (3). From this point the circle begins to get smaller and then eventually it disappears entirely.

Then, in the very next instant, the sphere disappears entirely from Flatland, having passed beyond the surface plane of the tabletop (or pond surface). From the Flatlander's two-dimensional perspective, the sphere has mysteriously disappeared. To a three-dimensional viewer, it is still there, but it has simply moved beyond the flat plane of Flatland.

Ours Is a Flatlander's Experience

From the Flatlander's perspective, this "alien" visitor to Flatland displayed very unusual behavior, and in their attempts to describe this behavior, the Flatlanders had to rely upon the concept of "time." They said things like *"it grew to a large size in one minute, and then shrank just as quickly before completely disappearing."* This strange and curious object broke all the normal rules. It came from nowhere and disappeared just as mysteriously, and while it was visible, even though it was a circle, it continuously changed in size. What was this mysterious object that grew and shrunk over time? What happened to it? How did it become invisible?

Describing the sphere and its behavior would be especially confusing for Flatlanders because circles don't normally change size; from their perspective this sphere must have been a magical or paranormal object not obeying any of their logical "rules." The lead character in this novel was ridiculed and perceived as crazy when he tried to describe to others what he had witnessed.

From our three-dimensional perspective there is no mystery because the physics and geometry of a sphere intersecting a flat surface are fully understandable. From our perspective is not strange or paranormal; it can be explained with simple geometry. However, this geometry requires the use of three dimensions, which is outside the realm of Flatland. Our native ability to perceive in three dimensions is a type of awareness that is beyond a Flatlander's *conceptual horizon.*

This story illustrates how, within our three-dimensional world, phenomena that appear mysterious, or seem to be changing in very strange ways over "time," could become quite easy to visualize and understand if we were able to understand and employ an additional dimension. This is a key idea to grasp, so I will repeat it in a slightly different way: ***strange or unexplained phenomena, and even our understanding of the way things age or change with "time," is nothing more than our limited way of glimpsing or understanding the geometry of extra-dimensional space.*** As we will learn, this is exactly the type of relationship Einstein illuminated in his Theory of General Relativity when he demonstrated that gravity was actually a deformation of four-dimensional *spacetime.*

This Flatland exercise demonstrates the similar difficulty we have when interacting with four-dimensional things from our three-dimensional reference frame. We only observe odd-looking pieces, artifacts or partial views of these objects; no human can really understand a four-dimensional object from our three-dimensional perspective. In fact, as illustrated in Flatland and our own lives, we are not even experiencing the full range or extent of all of the objects' three-dimensional qualities. Like the residents of Flatland, who lack depth perception, we ae missing the senses to perceive many things that exist within our own three-dimensional space; high and low frequency colors and sounds are only the most obvious.

If Flatlanders could suddenly understand three dimensions, then at that moment, their entire view of the universe would dramatically change. Of their many potential discoveries, two are of particular importance for understanding this book. *First, they would discover that they did not understand the full extent of many of the objects that are in their own dimensional realm. They would realize that even two-dimensional objects simply cannot be fully understood or described using only their two-dimensional concepts and mindset.* For example, even though the entire triangle is within their world, they never actually "see" the full shape of a triangle: they only observe a line that changes length. Two–dimensional beings cannot even experience the full range of two-dimensional shapes. Because objects can only be accurately described if we also understand the space that contains them, we require three dimensions to fully describe two-dimensional objects or two-dimensional space. When Flatlanders attempt to describe three-dimensional objects from their limited perspective, they automatically introduce a great amount of complexity, mystery and confusion.

The obvious implication is that, as three-dimensional beings, we cannot fully see or understand all aspects of even three-dimensional shapes. How then could we hope to understand or explain four-dimensional things? Because of this "dimensional problem," our attempts to describe four-dimensional space and objects often sound very complex and confusing.

The second critical discovery would be about the role of time and feelings in their "universe." *In the dimensionally restricted Flatland, time and feeling are clues or "windows" into the deeper structure of the next dimension!* Like Flatlanders, we also only interact with that limited part of creation that is accessible from our dimensional realm. Because of our built-in limitations, we also describe many phenomena using time and feeling, and are not aware that we are really trying to describe a geometry that lies beyond our ability to understand. It is likely that, for us, time and feeling provide windows that allow small glimpses into the next dimension.

I have been speaking of "objects," but all of this also applies to people, ideas, energy, and everything that is part of our three-dimensional reference frame. ***Behavior and phenomena, which seem complex or strange from our limited perspective, become clear and even simple if observed from an expanded, dimensional reference frame.***

VIEWPORTS

Throughout this book. I will use the term *viewport*, a descriptive term borrowed from a software program that I rely upon in my architectural practice. In this program, *viewport* refers to a special presentation "window" used to hide certain elements in a drawing, so that the drawing becomes easier to understand. Today's architectural drawings are typically constructed in three-dimensions, but showing all three dimensions can sometimes be confusing because there is too much *information*; the amount of *information* can be so overwhelming that the important parts become obscured. The *viewport* is used to simplify and show only the *information* necessary for the presentation. It can be adjusted so that only two dimensions are visible at any given time, hiding any unnecessary notations. This ensures a simple and direct presentation of the most critical *information*. The other *information* is not gone or lost; it is simply no longer visible. ***The extra information is still present, but it is now invisible because it has been placed beyond the viewer's senses.*** The software settings control what my clients can see.

This is a great metaphor for describing our three-dimensional experience within a multi-dimensional universe. We only have a limited *viewport* of the much fuller universe. Incapable of understanding and processing all of the information, we clearly "see" only three dimensions, and in addition we have a misty window into the fourth, which we call "time." The parameters (or settings) of our *viewport* will determine our experience of life. ***To help us function efficiently in our three-dimensional world, our viewport intentionally filters out the non-critical or extraneous information.***

The physical size of our *viewport*, our visible universe, is presently limited to about 15 billion *light-years* in any direction. This is the outer limit of what we can presently observe with our astronomical tools and, not coincidentally, this is also the distance that light has traveled since the cosmological event named the *Big Bang.* Our *viewport* is also restricted in size to things that are within the range of our senses and our tools. If we were much smaller or larger, our direct observations would be very different. Additionally, our *viewport* is shaped by the way our brains filter and process information—filters such as *cognitive dissonance* and the limitations of our five recognized senses. All of these factors work together to limit the quantity and quality of the *information* that we consciously process. These "settings" always form and shape our "point of view."

Of course, our *viewport* has both a conscious and a subconscious component. Just how big and important is the role of our subconscious to the whole of our expressed being? One way of measuring its effect is to look at how much relative information our conscious and unconscious minds can process in a given amount of time. According to one recent scientific study, our conscious minds can process between twenty and forty bits of information in one second. In that same second, our subconscious minds will process almost 20,000,000 bits of information. Since the subconscious can process up to a million times more information, all of it hidden from our conscious minds, we tend to be on autopilot all the time, without ever realizing it.

The subconscious is big, fast, and powerful. However, because its world is hidden from our conscious minds, its activity is not easy to self-monitor or access. Due to the preset parameters of our *viewport,* we consciously interact with very little of that which lies "beyond" or "deeper within," yet we are still fully connected below the surface. As a result, even though we see and consciously participate in only a very limited slice of creation, we are still a communicative and interactive part of a much larger organism! Later we will discuss why "beyond" and "deeper-within" become equivalent ideas, once they are explored from a more expansive and *holographic viewport.*

We can define our "viewport" as the "window" through which we on Earth consciously experience our particular slice of the universe. The precise extents of personal *viewports* will be slightly different for each individual, but generally all of our individual human *viewports* will fall within a range of common and expected experiences.

Einstein once described the nature of our *viewport* this way: "*Nature shows us only the tail of the lion, but I do not doubt that the lion belongs to it even though he cannot at once reveal himself because of his enormous size."*[1]

PROJECTIONS

When the three-dimensional sphere passed through Flatland, it appeared to be a two-dimensional line, which changed in length as time passed. These dimensional relationships can be described through a common mathematical function called a *projection.* Understanding this concept is important for understanding many of the ideas within this book. Everyday examples of *projections* include the images we see in movies, mirrors, maps, photographs, and shadows. With these, what we experience is not the actual object; instead, we interact with an image of the original object. For movies, we even use the term *projection* to describe this process. Actors play their roles in three dimensions, but their performance is reduced to two dimensions, both on the film and on the screen.

In Flatland, the actual object visiting was a sphere; but because their *viewport* was limited to only two dimensions, it appeared to the Flatlanders as a circle that changed size with time. This circle was the *projection* that the three-dimensional sphere "cast" onto the two-dimensional plane of Flatland.

---

[1] As quoted by Abraham Pais in *Subtle is the Lord: The Science and Life of Albert Einstein* (1982).

If we were to stand outside and look at our shadow, we would be looking at the *projection* of our standing, three-dimensional body onto the flat, two-dimensional plane of the ground. This *projection* or shadow is not the object; it is the reduced or *flattened image* of the object that has been transferred, or *mapped,* onto another system of *coordinates.* In this case, this "other system" is the flat ground plane.

A shadow lying entirely on the flat plane of the surface of the Earth is a 2D image or projection of a 3D object.

If we were the size of bugs, living entirely on the flat surface of the Earth, unable to look up or down, much of our awareness of people would be limited to encountering their shadows. From a bug's perspective, imagine how very difficult and mysterious the understanding of humans would be. How could a bug even begin to describe us from their encounter with our shadows?

On Earth, we see and understand everything through our three-dimensional *viewport,* even though, as we will discover, the universe is built from at least four (and likely many more) dimensions. **What we observe and experience on Earth is only the very limited projection of our full, multi-dimensional being onto a three dimensional space. We experience only the shadows that intersect, interact, or map onto our three-dimensional world—a greatly reduced "image" of the full thing.**

Throughout this book, I will use the terms *projection, shadow, image, illusion,* and *dreamscape* to describe this phenomenon. *Projection* and *image* are the mathematical terms, while *shadow* more accurately describes our sensory awareness. *Dreamscape* and *illusion* more closely relate to the psychological aspects.

Physicists have recently been referring to the term *hologram* to describe the *image* quality of our physical world. This is an excellent metaphor and later we will discuss *holograms* in depth. We will also learn how *holograms* are even more interesting because of the way *information* may be stored throughout our universe.

Projections Create Odd-looking Artifacts

As we explore the concepts and worlds that unfold through extra dimensions, we always need to keep in mind a very important principle—**whenever we interact with or view something in fewer dimensions than its actual full geometry, what we encounter is only a slice of the full expression of the actual object, idea, or experience.** As discussed, we are not capable of visualizing or fully understanding a multi-dimensional object, experience, or concept because our input and processing is limited by the extents of our *viewport.* We consciously make contact with only the *projections* of the *projections,* or the shadows of the shadows, which are *projected* or cast from the original objects, and filtered as their images pass through multiple dimensional realms before eventually intersecting our world. What we see and encounter may not even resemble the real thing because, through this process of *dimensional reduction*, the original information is distilled and distorted many times. Our *viewport* only allows us to experience those shadow aspects that are understandable by us. **"Dimensional reduction" is my descriptive name for this reductive process, involving multiple levels of projections. This "dimensional filter" is the primary reason for the appearance of mystery in our lives.**

Dimensional reduction plays tricks on our sensory systems, specifically when it involves our ideas of "time" and space. Later we will discuss Einstein's realization that the *force* we call *gravity* is only our limited experience of something occurring in four-dimensional space; the *force* that we eventually experience is only an artifact.

For this book to make complete sense, we need to understand the very similar concepts of *dimensional reduction* and *dimensional filtering*. The simple and common task of making a map can help us better understand what these ideas represent. (Note that these same terms also have very specific technical definitions that are quite different from their use in this book.)

MAPMAKING

Mapmaking is useful for demonstrating how *dimensional reduction* causes a profound loss of information. Mapmaking involves taking a three-dimensional object, such as part of the surface of our curved Earth, and adjusting its information so it fits within a flat, two-dimensional representation called a map. To accomplish this mathematically we use *algorithms* or formulas. Since a two-dimensional map is displayed on a flat piece of paper (or smartphone screen), essentially we are eliminating the third dimension by flattening the Earth's curve. The larger the piece of the Earth that is mapped, the greater the information loss and distortion.

In other words, when we use a map, we are viewing a three-dimensional object through only two dimensions; we must therefore lose or distort the *information* about the third dimension. There is no accurate way to express the entire spherical Earth on a flat piece of paper. *Information* about the third dimension first needs to be *flattened,* a form of filtering, before it can be presented on flat paper, a two-dimensional *viewport*.

One of the most common methods for making a world map is the *Mercator Projection,* and the most common version of this is shown below. This particular *algorithm* is fairly accurate near the equator; but as you move to the poles, the land masses and distances get very distorted. Canada and Greenland seem to be enormous landmasses, but this distortion is only an artifact from this process of *dimensional reduction*.

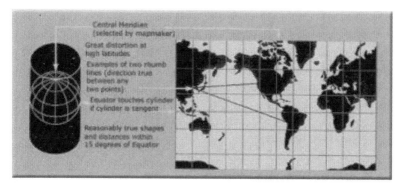

In a Mercator Projection, each rectangle represents an equal area between the lines of latitude and longitude. The areas near the equator are the most accurate. There are other versions of the Mercator that create different areas of distortion.

Hundreds of different types of popular *map projections* can be constructed. While all *projections,* or two-dimensional maps, of our three-dimensional world are inherently inaccurate, each is useful for its own reasons. Inaccuracies will always arise when *information* is lost or altered through the elimination of even a single dimension.

Our World is Similar to a Map Projection

Imagine what could change and what inaccuracies would occur if this process of *projection* and *dimensional reduction* happens at least seven or eight times in succession. This occurs when *information* that originates in ten- or eleven-dimensional space is *filtered* or *projected* through each level until it eventually reaches our three-dimensional universe. What we ultimately experience is very different from the original *information* at its source.

We can think of our world as a three-dimensional "map" of a topology that has many more dimensions, which have been reduced and crammed into just three dimensions. Due to this reduction, whatever our experience is, it will not be an accurate description of what is really happening within the larger dimensional spaces. Our conceptual minds only understand the geometry and limitations of our own *viewport.* Just as with the software program, the original *information* is still available, but we are unable to see it directly: it is hidden because of our *viewport.* We only directly experience the shadows or *projected image.*

In addition, because we have our own personal filters, we each unconsciously modify this cultural *viewport* to create our own specific personal *projection*. Through living and personal growth, each one of us is tweaking the *algorithm* that defines our personal *viewport,* and therefore **we each see our lives through an individual and unique reference frame.**

Assuming that we all share exactly the same *viewport* or worldview has been a common source of many of mankind's problems and conflicts. Two different people can witness the same event—a car accident, for example—and their reports will often read as if they had been at two entirely different places. It is not that one view is right and the other view is wrong. Since we each unconsciously frame our *viewport* in a slightly different and personal way, we all perceive and cognize *information* uniquely.

If we allowed more room for the unique framing of *viewports* by different individuals and cultures, the blame and divisive idea of "right and wrong" could begin to evaporate. ***It is, therefore, important to realize that every one of us is experiencing our own reality through a slightly different and uniquely personal projection. We all live in slightly different realities, so our experiences are always ours, and ours alone.***

VIEWPORTS SHIFT OVER TIME

Our individual and consensus *viewports* constantly shift, change shape or size, and evolve. When we observe an old paradigm changing around concepts such as flat Earth, apartheid, women's rights or civil rights, what we are actually experiencing is a gradual shifting of the collective *viewport*. However, this is a difficult process and the amount of time required to make these collective, cultural shifts often feels painfully slow. Fortunately, waiting for others is not necessary; individuals can shift their personal *viewports* much more rapidly (even instantly) than an entire culture; and once this happens, all those "others" interacting with the shifted individual will also appear to have changed. ***Becoming aware of our personal power, through our ability to shift our own viewport, is an enormous step towards personal freedom.***

VIEWPORTS IN THE *WEB OF POSSIBILITIES*

No human *being* is capable of grasping the complete multi-dimensional picture of "all that is," but we each can still recognize our deep and fully interconnected relationship to this more expansive structure, and live accordingly. We all have occasional hints, revelations, quick glances, or moments of knowing that grant us glimpses beyond our normal *viewport.*: these are our invitations to recognize the broader perspective. If we are open to this guidance, we can begin to make lifestyle changes that ultimately result in our ability to inhabit a different and larger landscape within the "Web of Infinite Possibilities."

My description of the Web includes the word "possibility," which refers to a specific concept from *quantum physics*. As we will soon discuss, years of research have determined that, at the particle level, there is a quantifiable *probability* for every possibility. Since we are built from these particles, anything is also possible for us—including the possibility of completely changing the world that we experience. This is wild stuff but, if we follow the physics and examine our actual history and experiences, this idea actually begins to make sense.

Once our personal viewport shifts, even slightly, it will seem, at least for all practical purposes, as if the entire "outside" world has completely changed. When this happens, the "outside" world really has not changed at all; what has changed is our own position or perspective within the continuous Web. This new perspective is the result of altering the size and extent of our personal *viewport.*

Typically, from our new *viewports,* some things will persist unchanged. Most of our friends will remain, our home is usually unchanged, and we may still work for the same company. We often need to be unusually observant to even register the subtle changes; but as we observe the details of our life more carefully, we may discover that enough has changed to effectively alter our relationship to the outside world! Maybe our smile is met with a warm smile from a once-distant co-worker, our boss notices our extra work on a particular project, the dog's barking didn't keep us up last night, and that tired feeling is gone, replaced by the renewed excitement around some new endeavor or relationship that unexpectedly came our way.

These are small and ordinary changes: common everyday events that can easily be overlooked. Incremental, small shifts like these are actually best for personal change because we frequently have difficulty adjusting to larger and more dramatic shifts. ***Over time, the accumulation of many of these tiny shifts results in a very different personal adventure.*** Small changes continue to occur freely, moment by moment; and one day we wake up and realize that we are actually living a completely different life, one that is likely to be much more harmonious, joyful, and more in-tune with the life that we imagined possible. This is the universal observation of individuals who have committed to this lifelong process of growth, change and opening. "Shift happens" when we open to these new possibilities.

All shifts to our position in the Web are the result of a change in the way we vibrate at the core of our being. As we allow for the expansion of our awareness, our vibration changes and our *viewport* becomes larger. The result of this type of growth is that we have a richer experience and relate to more of the "outside" world. Life usually becomes much more vivid and interesting.

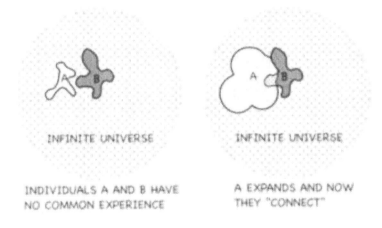

INFINITE UNIVERSE

INFINITE UNIVERSE

INDIVIDUALS A AND B HAVE
NO COMMON EXPERIENCE

A EXPANDS AND NOW
THEY "CONNECT"

Individual "A" has shifted and expanded his or her awareness. He or she now shares common ground with "B." Where they once were unaware of each other and had very separate lives, they now experience connection. The "outside" world did not change, nor did individual B. We only need to change ourselves to change our interaction with "others."

Understanding quantum mechanics and the Web of Possibilities is not a necessary requirement for creating or experiencing personal shift. We cannot accomplish this type of change with our minds alone;

only resonant change at the deepest levels of our being can initiate this kind of shift.

Eckhart Tolle summarizes his perspective on shifting *viewports* when he writes, "When you are transformed, your whole world is transformed, because the world is only a reflection."

The *Web of Possibilities* exists now and forever, and yet always outside of time; but because of our special relationship to "time," the Web appears to us to be dynamic, growing, ever changing and evolving. Through practice and evolution, one day we will be able to move more consciously around the Web; ultimately, we may even find ourselves free to journey wherever we choose. For the present, however, we can appreciate and build upon the simple experience of witnessing or observing our everyday, small, personal shifts.

THE PHYSICS OF THE LAST 100 YEARS

INTRODUCTION

This section covers the history and meaning of the most relevant underlying physics. Many of the ideas in this book began with my study of physics, and then later were refined through personal experience and my study of historical and spiritual traditions. I have tried to present the physics with historical context and without technical detail so that readers with no scientific background can more easily follow. While some readers may reflexively shy away from any discussion involving math, physics or science, they need not fear this particular chapter. This presentation involves no math or manipulation of equations; it is simply a historical story with a few, scattered, personal insights, experiences, and anecdotes.

At the same time, please do not get bogged-down in this section. **I ask all readers to give this science section a try, but if it proves to be too difficult or not of interest, just skim or skip it. The book's main ideas can be understood without ever reading this section; and, if desired, this section can always be read later.** If, on the other hand, a reader desires a deeper exploration of the physics, *The Architecture of Freedom* covers physics in more detail; and both books include references and sources for deeper exploration.

Knowing something about the science will help readers to better understand the ideas and architecture that I am describing. This book will only make sense once the reader accepts that **the universe is infinitely larger, deeper, and more interconnected than we generally imagine; and much of what we consider to be the limits or bounds of our existence are simply the by-product of our type of conceptual thinking, our limited senses and tools, and our culture.** Once we fully understand this principle, the rest of the book should flow and make complete sense.

Having studied only undergraduate level physics and teaching a couple of years of high school physics, I am far from an expert in *relativity* or *quantum physics*. I do not live and breathe the theory and mathematics like many practicing physicists, so I must trust the experimental analysis and more technical aspects to those who are fully immersed in the science. My ultimate interest has always been to understand just what these discoveries mean for our day-to-day lives.

Freed from the tedious calculations and dependence upon grants and professional peer review, I have been mostly like a kid in a playground, joyfully exploring what is possible while using my friends and myself as the experimental subjects. It was largely through this type of "play" that I began to discover the more practical applications of this physics. This playful beginning has evolved into a heartfelt vision of a complete universe with clear purpose and meaning. Infused into this is an awareness of the infinite possibilities for this wonderful adventure that we call life.

THREE SETS OF PHYSICS

Physics is the most basic scientific study of how physical things work in our world. Today's physics is rapidly evolving and at a very exciting threshold but it still finds itself divided into three somewhat independent sets of theories or laws.

First, we still use the extremely functional set of laws derived from Newton. This is the physics that we now call *classical physics;* it includes Newton's laws of motion and all the additional physics that they have spawned.

There is also a set of rules to describe very large things, high speeds, and great distances, such as *galaxies, light,* and *gravity.* This physics is built upon Einstein's *relativity. Relativity* can be understood as a revision or addition to *classical physics,* meaning most physicists see our physics divided into only two different parts—*quantum physics* and a combination of *classical* and *relativity.* Since 250 years and a paradigm shift actually separate *relativity* and *classical physics,* I will continue to treat them as distinct and separate branches in this book.

Third, one set of physics describes very small things such as atoms, subatomic particles, and the *energy* and *forces* that are associated with them; this realm of the very small is called *quantum physics.*

A primary focus of contemporary physics involves trying to integrate these somewhat separate descriptions into a single theory, which, alone, would describe the behavior of everything in the universe: the *unified field theory. Unification* is the greatest problem of modern physics, but so far our attempts at solving it just raise new questions that all seem to point towards a deeper, but well hidden, truth.

Classical physics, which still accurately describes all that is found in the middle between the very big and the very small, is built upon that which we can see, experience and measure directly in our three-dimensional "solid" world. For almost 400 years, this physics made wonderful, logical sense to our rational minds; it is derived from the direct observation of our three-dimensional world, and it was developed through repeatable experiments that historically did not rely on extremely complex or expensive equipment.

The two newer branches, *quantum physics,* describing the workings of the very small *subatomic particles,* and *relativity,* seeking to understand the mechanics of the vast cosmos, illuminate many ideas that do not make the same type of "good sense" to our three-dimensional, and thus specialized, minds. Even though *relativity* often seems strange, we can think of this branch as an improvement or update of *classical physics* since they fit together mathematically. However, when discussing *relativity* and its meaning, we also find that we must venture outside of our long-understood, familiar and "safe" three-dimensional paradigm.

Even though it has been unbelievably successful and thoroughly tested for almost a century, *quantum physics* does not seem to integrate mathematically with the other two branches. *Classical physics* is a study of exactness and certainty, while q*uantum physics* is about a world of possibilities and probability. To our three-dimensional and logical minds, these two approaches make little or no sense together.

As mentioned above, the advances in *quantum* and *relativistic physics* have directly led to the development and recent deployment of an abundance of new, powerful, and very functional technical devices. We have so fully integrated these machines and tools into our contemporary lives that not only are we are entirely dependent upon them, but we even take them for granted. If most of us actually understood the real science behind our computers, iPods, automobiles, electrical grids, GPS systems, weather satellites, phone systems, or weapons, we would be in constant awe and perpetual recognition of our unique place in the history of this rapidly changing paradigm. The strange mathematical predictions, the subsequent exploration of the behavior of subatomic particles, and our deepening knowledge of the cosmos have led directly to the development of this wide array of electronic gadgetry that generates, drives and moves the vast amounts of *information* that today's world depends upon. We often hear that "the proof is in the pudding," and it seems that we now have more than enough pudding.

Quantum theory radically changes how we must look at our world, and if were simply an untested theory, then this entire book could be described as science fiction. However, every test of the quantum theory during its almost 100 years of existence has resulted in its absolute confirmation; no other major theory in physics has ever had this success rate. As much as one-third of our national Gross Domestic Product (GDP) is dependent upon or built from products based upon this theory. The high-tech products from this theory have become as much a part of America as apple pie. Quantum theory, which has become a very critical and integrated part of our lives, is at least as real as anything else is in our world!

Today, we use these three sets of rules and observations to describe everything that we understand about our universe. *Classical physics*, formalized by a young Sir Isaac Newton more than 400 years ago, works best with objects sized for the human scale: things such as baseballs, bugs, airplanes and bridges. It is relatively simple, intuitive, and practical—most of us have a gut-level understanding for much of this physics, even if the math looks complicated. We know that if we throw a ball into the air it will come down and if we hit a brick wall we will quickly stop moving. *Relativity*, Einstein's contribution that describes the very big objects we encounter at the cosmological scale provides us with the tools to begin understanding and describing things like *gravity*, the speed of light, and our galaxy's origin in the *big bang*. Much of *relativity*, however, is counter-intuitive to our three-dimensional mindset, so its incorporation requires hard work and a great deal of imagination. **On the other hand, quantum physics, the study of the very tiny, can only be described as completely "weird and crazy." Understanding its implications requires adopting an entirely new perspective on the nature of the universe, and even on life itself.**

The fact that we still need more than one type of physics to fully describe our world is a clear sign to most people, and to virtually all physicists, that we do not yet really understand our universe. Most physicists believe in the existence of a single, final, *unified field theory* that would incorporate and connect everything that we know to be true: a theory that would *unify* the four known *forces* (*gravity, electromagnetic, strong nuclear* and *weak nuclear*) into a single elegant theory. The hope is that, when found, it could be expressed as a simple equation or a set of equations describing everything known.

To date, this *unification* has completely eluded the efforts of our very best scientific minds. As a result, many scientists acknowledge a general level of discomfort as we continue to divide the physical world into three somewhat disconnected parts. A *unified field theory* would change this but, in the effort to reach this important milestone, many scientists have devoted entire lifetimes without a clear breakthrough—Einstein spent most of his life and career trying! His three major theories, *special relativity, photoelectric effect,* and *general relativity*, came very early in his career; much of the rest of his life was devoted to his struggles with *quantum theory* and the *unification* of the four known *forces*.

The problem is that neither *quantum physics* nor *relativity* can be explained or understood within the framework of our conceptual thinking or through the older, three-dimensional *classical physics*. We are consistently uncovering concepts that are so unexpected and profound that they are impossible to understand or imagine, even for the most talented physicists. **For all of us, born into a world that was carved and framed by rationality and this classical physics mindset, to even begin incorporating the meaning of the last hundred years of research requires a complete paradigm shift.**

Remember that it was not too long ago—less than 500 years—when virtually all humans thought the world was flat and ended with a distinct edge over which one could fall. Back then, all people believed that the Sun revolved around the Earth, which was presumed to be the physical center of absolutely everything in the universe. The Earth was the most important part of the universe: the focal center for all of creation. From within that old, rigid mindset, it was impossible for most humans to imagine the next leap, which was the understanding and cultural integration of the real geometry of our solar system, a structure that Copernicus discovered and Galileo later refined.

At that time, this change of vision required an enormous conceptual leap: the understanding and acceptance that our planet was just one part of a much larger system, the solar system, with our Sun as the central element. 400 years ago, we began our shift to the *Heliocentric* (Sun-centered) view, as we gradually abandoned the old *Geocentric* (Earth-centered) view. Through the years, we have further refined this new paradigm by discovering that our solar system was still within a much larger system called the *Milky Way Galaxy*. We realized that even our local galaxy, which contains some two hundred billion stars, is just a small part of a much larger physical universe, containing, at least hundreds of billions of other *galaxies* of similar or greater size than the *Milky Way*. Our understanding of the vastness of our physical universe has so shifted the cultural understanding that it has now become amusing to imagine that people once thought that the Earth was flat and situated at the center of the entire universe. ***Humans have a long history of successfully vaulting through dramatic paradigm shifts. We are on the verge of doing this once again.***

Throughout this human journey, physics and math have always been our harbingers, pointing towards new directions for exploration by regularly providing the first insights and clues. Today, looking back we clearly see this repeating pattern of successfully following the best math, and then science, deep into what was once unknown and strange territory. Eventually, as this once-strange science becomes more integrated, it begins to seem quite normal. ***As a culture, we have shown that we can and do dramatically change the way we think; but it always involves a complex and extended process—one that has been successfully guided by our best theoretical math and science.***

THE NEED FOR EXTRA DIMENSIONS

There is general agreement among many physicists that we are much closer to uncovering this "Holy Grail of physics"—the *Unified Field Theory:* the single theory that combines *quantum mechanics, classical physics,* and *relativity*. Many physicists are very excited about the newest modifications to *string theory* and their possible role in this *unification.* Now forty years old, *string theory* is a more recent and mature sub-theory of *quantum physics.* Working with the extra seven or eight extra dimensions required by *string theory* is relatively easy using the tools of mathematics. Once we add these dimensions, the three sets of physical laws begin to merge, and mathematically work together in a more elegant way. However, at the same time these added dimensions are also impossible for any of us to visualize or really understand.

In this book, I explore several different independent lines of reasoning that all lead to the same conclusion that our universe is built upon extra dimensions. The addition of extra dimensions changes everything that we understand about our universe. It takes our letting go of old ideas and concepts to begin to grasp what a universe constructed this way might mean. Before we wade into these discussions, let us first back up and examine the history of modern physics more closely.

CLASSICAL PHYSICS

Our Ancient Fathers

The ancient Greeks were largely responsible for embedding the Earth-centered model of the universe into our consciousness; they gave us two of our longest lasting paradigms: Euclidian three-dimensional geometry and the "flat earth" model of the universe.

Euclid, who walked ancient Greece about 300 BC, is the father of our three-dimensional mathematical language, and today we honor his contribution by calling this mathematics *Euclidean geometry*. His contemporary, Aristotle, was the first person of record to declare that there are only three dimensions to the universe. He said, "The line has a magnitude in one way, a plane in two ways, and the solid in three ways, and beyond these there is no other magnitude because three are all." Several hundred years later, Ptolemy proposed a mathematical proof of only three dimensions, based upon not being able to draw a single line that is perpendicular to all three axes of a three-dimensional object. For a long time this seemed like the end of the story; it had been declared by the finest minds that *"three are all!"* For the next two thousand years, we found ourselves deeply locked into this geometry, one that is easily understood by our senses and logical minds. Except for a few isolated and almost forgotten "fringe-thinkers," no one seriously imagined that we were missing a big piece of the picture until the middle of the 18th century. "Curved" or multi-dimensional space was not described, documented or even thought possible until about 150 years ago.

The ancient Greeks were also largely responsible for embedding the geocentric (Earth-centered) model of the universe into our consciousness; they gave us two of our longest lasting paradigms: Euclidian 3D geometry and the "flat earth" model of the universe.

Physics Between the Times of Newton and Einstein

Through the Dark and Middle Ages of humanity, there was little time or energy for pure, rational science. Instead, through trial and error people learned how to strengthen castle walls, build higher cathedrals, make stronger armor and swords, and develop more effective and lethal weapons for warfare. Few, perhaps even none, of these activities were approached scientifically, and the church often crushed any real attempts at scientific progress.

Starting in the late 1600s, our Western culture began to be shaped by the newer mindset based upon Newton's description of our physical world. With his *laws of motion*, we entered into a time when man could imagine his world as a mechanical machine governed only by these laws. There was the common and popular belief amongst scientists and philosophers that, since many physical observations could be analyzed, predicted and repeated, eventually this logical system would unlock a world where everything could and would be predictable. The rational philosophy of Descartes meshed perfectly with this new attitude, and together they intertwined to shape a brand new paradigm. God still existed, but many people began to believe that his job description was reduced to "just controlling the gears." Anomalies, whenever they popped up, were usually dismissed as poor or inaccurate science. It was believed that absolutely everything could be figured out within these new principles because everything was predictable except when God occasionally intervened by adjusting the controls. Newton's declaration that "objects at rest tend to stay at rest, and objects in motion stay in motion unless acted upon by an outside force," along with his famous companion equation, force equals mass times acceleration (F=ma), became the unmovable foundation upon which this new paradigm was built.

For well over 300 years, *classical Newtonian physics* seemed to have the nearly flawless ability to answer almost every question thrown in its direction. Then, just before the time of Einstein, we developed new scientific and mathematical tools as the Newtonian model started to show its extremely worrisome and paradigm-threatening faults. It quickly became clear that some very large pieces of the puzzle were still missing. To better understand the context of this change, one that we are still embroiled with, let us next examine the history of our exploration of *matter* and the cosmos.

Size of the Atom

Ancient Greek Philosophers predicted the atom as the elemental building block of matter, but it was not until the 17th and 18th centuries that we began to understand the actual chemistry of the atom. The atom is extremely small; a half-trillion of them (500 billion), packed tightly edge to edge, would form a line only one inch long. If you were to take a marble and blow it up to the size of the Earth, only then you would be able to clearly see the atoms inside it because, at that magnification, atoms would become about the size of marbles. Stated in a slightly different way, the size of an atom, compared to the size of a marble, is the same as the size of a marble compared to the size of the Earth.

Today we understand that atoms are far from "solid." They mostly contain empty space, with smaller *particles* (*protons, neutrons,* and *electrons*) that help define the borders of their space. How much empty space is in each atom? The nucleus (*protons and neutrons*) of an atom has a diameter more than 10,000 times smaller than the atom itself. The only other "things" in the atom are electrons, which are so small and flighty that they do not take up any real space at all. This means that atoms are almost entirely empty space.

A few additional analogies can give us a better sense of how much empty space is inside an atom. If we inflated the smallest atom, the hydrogen atom, to the size of a football stadium, the nucleus would be the size of a pea in the middle of the fifty-yard line, and the single electron of hydrogen would be smaller than a speck of dust in the stands. The only actual matter in a region the size of a large football stadium is about the size of a pea. ***Empty space makes up the vast majority of the solid appearing atom.***

The atom certainly is not the solid thing we experience when, for example, we hit a brick wall with our fist. As we peered into the nucleus to study protons and neutrons, we discovered that these very small "particles" that make up the only "solid" parts of these atoms are similarly constructed—they also follow the same "mostly empty space" pattern. As we continue to discover new ways to peer into the subatomic world and off into deep space, it seems that this pattern repeats itself, on and on in both directions, towards both the smaller and the larger.

As we explore this topic in more detail, we discover that this "empty" space itself is not really empty at all. Data from very recent space probes indicate that "empty" space contains most of the gravitational material in the universe, and may actually hold the key to the ultimate destiny of our physical, three-dimensional universe.

Our World Is Not So Solid

Since the atom is mostly "empty" space and everything that we interact with in our day-to-day world is made of atoms, if we were small enough—say, the size of a *nucleus*—we would be able to see and directly experience all of the space that is within and between the atoms and understand that "solid" things are not actually very solid after all. They just seem solid to our senses from our particular perspective. ***If we were the size of a nucleus or smaller, looking at the space between the particles would feel much the same as peering at the space between distant stars.***

Today, we define four primary *forces*. The *electrons* and *protons* are all held together by *electromagnetic force*. The *strong nuclear force* holds the similarly charged and therefore repelling particles of the *nucleus* together. Along with these two forces, we have also identified the *weak nuclear force* and *gravity*. Einstein demonstrated that our experience of *gravity* as a *force* was only an artifact created by the extra-dimensional geometry of *spacetime*. It seems likely that we will eventually understand these other *forces* in a similar way.

What we perceive as "solid" material is mostly space and energy (attractive and repulsive forces), with a microscopic amount of solid-appearing particles. ***Our bodies are made up of atoms, so we, too, are constructed from space and energy. Everything that we call matter, including our bodies, is really only organized energy.*** As we begin to understand this concept, the expected outcome from two "solid" appearing objects bumping into each other becomes less certain and less predictable. At some point, it becomes almost easy to imagine how the tiny "particles" of one object could pass through the enormous voids of the other. If the *energetic* conditions were right, our bodies should be able to easily pass through walls.

Euclidean Geometry Has its Limits

Euclidean geometry is what we all instinctively rely upon to find our friend's house, fit furniture into our rooms, throw a baseball, read a map, build a wedding cake, and determine if we can safely jump to the next rock. Even if someone has failed high-school geometry, they will still have a deep understanding of this geometry; it is built so deeply into our bodies and brains that, even though we rely on it constantly, we never have to even think about the math. However, this every-day, three-dimensional geometry, which has served us so well and for so long, has now been discovered to be incomplete, inaccurate and, possibly, even wrong in certain situations.

Classical Physics Falls Short

By the turn of the 20th century, fresh new ideas and strange experimental results began to raise clear and deep questions about the infallibility of the classical mechanistic model. What exactly is gravity? Why don't Newton's equations precisely describe the motion of the planets? What is light? Why don't electrons collapse into the nucleus of the atom? The physics that once promised to explain everything left these, and other, questions unresolved.

Relativity and *quantum physics* changed everything. Evolving independently, these two new theories allowed scientists to take a fresh look at these questions. As the years passed and these theories proved extremely successful, scientists and philosophers started to imagine that these theories must also be describing a radical new way of viewing our universe and life itself.

Understanding and fully integrating this growing body of revolutionary knowledge will ultimately require a greater adjustment than that of our last great cultural shift, from the Earth-centered to the sun-centered *viewport*. *Relativity* and *quantum physics* have allowed us unprecedented access to the very large, very fast, very distant, and very small parts of our universe. Newton's laws still work, but now we realize that they work only when limited to certain sets of conditions or reference frames involving a specific range of size, speed and time: objects of a size, *mass,* and *velocity* that are related to the scale of our own bodies. However, when we analyze things so small or fast that we cannot observe them directly, or things so large, distant, or energetic that we have no frame of reference, the 400-year-old Newtonian laws of physics are revealed to be only working approximations.

SPECIAL RELATIVITY

Of the four groundbreaking papers Einstein published in 1905, the third was named *On the Electrodynamics of Moving Bodies;* only later was it re-named *Special Relativity*. In one sense, this theory was a modification to the long-standing agreement within *classical physics* about what scientists and mathematicians call *reference frames*. It began by reinforcing the existing classical view that all uniform motion is relative to the motion of the observer and that there were, therefore, no privileged or special *reference frames*. It said that because everything in the universe is in constant motion, any reference point must also be in motion. Anything observed is, therefore, always relative to that motion. This part of his theory still described *classical physics*, and caused no major conflicts or problems.

Speed of Light

Einstein then introduced the entirely new idea that the speed of light was the same for all observers, regardless of their reference frame. Treating light uniquely was a ground-shaking idea and, with this new realization, much of our understanding about light, space, gravity and time was suddenly and dramatically transformed! Once we understood and analyzed his math, we then realized that all our observations and measurements of time and distance depended on how fast we were traveling.

It was necessary to add another dimension to our coordinate system to make sense of these unexpected predictions. Einstein showed that time was not separate from three-dimensional space and, instead, we exist in a unified *spacetime* coordinate system—a four-dimensional blend or continuum that combines three-dimensional space with time. ***Once Einstein deduced that we were living in this multi-dimensional combination of time and space, the real meaning of "time" changed forever. Culturally, we have yet to integrate this dramatic, now 110-year-old realization.***

This new theory predicted some very odd and unexpected behavior for light involving the concept of speed. The measurement of speed is, of course, completely linked to how we perceive time. This is why, in a very real sense, this theory completely changed our understanding of "time."

Imagine that both you and your friend are walking along together at five miles-per-hour. From experience, you would imagine that you and your friend could walk side by side on a trail and have a conversation. If your car is going 50 miles-per-hour and your friend's car pulls up, you expect to be able to shout to each other through the open windows.

Light is different; it does not travel in this customary way. At slower speeds, it still seems similar. Flying in an airplane at 500 miles-per-hour you might see another airplane out the window. If the planes fly closely enough and parallel, you can wave and even get a response from people in the other plane because the planes are moving together at a relatively low speed.

However, light obeys a unique set of speed rules. Einstein demonstrated that you can never catch up to a beam of light because no matter how fast you travel, light always moves away from you at the *speed of light,* which is about 186 thousand miles per second, or close to 670 million miles per hour. No matter how fast you traveled, you could never catch up to a light beam, for it would always be moving away from you at 670 million miles per hour. Light behaves very differently from phenomena such as sound, which you can catch up to and even physically pass. Jets do this often, creating a *sonic boom* as they zoom past their own moving sound wave.

Einstein's theory also predicted that if we were to travel at these high velocities, very unusual things would begin to happen to us as our speed increased. An *outside observer,* not going the speed of light, who happened to be watching as we moved at these very high velocities, would notice that we would start changing shape. To the eyes or instruments of this *outside observer, as we moved faster,* our body would flatten or compress in the direction of our movement. Our movements would slow down so much that we would appear to almost stop. It would seem like *Gravity* had become much stronger. Time would slow down, we would age more slowly, and, as we approached the speed of light, the watch on our wrist would almost stop. **But from the perspective of a passenger traveling with us, nothing would have changed. To someone in our own moving reference frame, including our self, the world would still look and feel normal.**

These observations are very strange when considered through the "common sense" of our conceptual minds, which are designed for efficient navigation through and around our three-dimensional world. Despite these strange-seeming results, this physics – Einstein's *relativity* – has proved to be sound and durable, having been thoroughly tested and explored for more than one hundred years. The more we discover, the more credible this theory becomes.

E=MC²

From Einstein's fourth publication in 1905, *Matter-Energy Equivalence,* we also learned that *mass* is really only *energy* that is being expressed in a different form. In this publication, he demonstrated that *energy* and *mass* are interchangeable forms of the same thing; *matter* is *energy* that has been "frozen" and stored in a different state, and it can be converted back to *energy* under the right conditions. In this "frozen" state, *matter* contains an enormous amount of *energy*. This relationship is described by his famous equation $E=MC^2$, or "Energy equals mass times the speed of light squared." Einstein determined that to calculate the amount of *energy* contained in a given amount of *mass,* one needs to multiply the amount of *mass* by the speed of light multiplied by the speed of light— a very big number multiplied by a very big number resulting in an enormous number. This means that a very small amount of *mass* contains an unbelievably large amount of *energy*. A handful of material under the right conditions can release enough *energy* to destroy a large city, exactly what occurred in Hiroshima and Nagasaki at the end of World War II.

One special condition, which allows *mass* to convert quickly to *energy* in a very dramatic way, is the aggregation of a *critical mass* of *fissionable material.* An uncontrolled *fission* process leads to a violent release of energy — this is the atomic bomb! By clearly and dramatically demonstrating how a small amount of *mass* produces an enormous release of *energy*, this weapon became the experimental proof of Einstein's theory. But this particular method of proof was also his lifelong regret. When this release of *energy* is slowed and controlled, as it is within *nuclear reactors,* this same *process* can be used to generate heat and electricity.

The "bomb" is experimental proof of E=MC2.
A small amount of material releases an
enormous amount of energy.

Today, we universally understand that *matter* is just another form of *energy*. The recognition that everything in the universe can be reduced to *energy* entirely reversed the old classical view that *matter* and *energy* were two very different things, and this leads us to one of this book's critical concepts: ***There are no actual "things" in the "physical universe." There is only energy in its various forms.***

62

Everything you have ever or will ever experience can be reduced to energy and energy exchange. Understanding this concept fully can help us begin to let go of our old concept of a rigid, solid, and fixed universe.

INTRODUCTION TO GENERAL RELATIVITY

Ten years after publishing *special relativity*, Einstein further modified this work with the publication of his *General Relativity* theory. With this update, there was finally an explanation for *gravity*, which had always been the "elephant in the room" of *classical physics*. Einstein showed that *gravity* is a temporary warp in the fabric of *spacetime,* caused by the presence of a large *mass*. This effectively replaced the idea developed by Newton that *gravity* was a *force* that causes masses to attract each other.

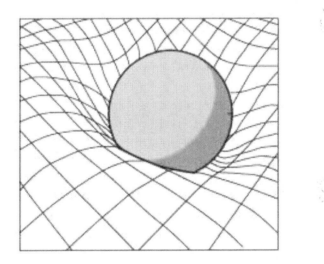

A heavy mass warps spacetime—a combination of 3D space and time— in much the same way the mass might warp the flat surface of a trampoline.

Suddenly, what was once described as a "mysterious" *force* could be seen as the direct result of an unseen but very real geometry. *Beginning with this realization, understanding the "mystery" became more about understanding the deeper geometry. Gravity, something we rely on every day and we all readily and intuitively "understand," turns out to be a function of the architecture – the geometric shape and structure – of our universe. We are not able to see this deformation of spacetime with our normal senses because we are equipped to see the world only in three dimensions. Instead, we "feel," and then interpret this four-dimensional deformation as a force called gravity.*

This is similar to our old understanding of "time" marching along a one-directional arrow. One day we will clearly realize that our understanding of "time" is only our limited perception of something that is an integral, geometric part of an extra-dimensional architecture.

WHAT RELATIVITY MEANS

Relativity signals the erosion of our existing cultural paradigm in multiple and dramatic ways. It demonstrates that we live in a space that has more than three dimensions: Einstein's *spacetime*. It describes how *spacetime* is actually "curved," and in local areas how this curvature creates the illusion of a *force* that we call *gravity*. *Relativity* demonstrates that *gravity* affects the rate of the unfolding of time, and we are only beginning to understand what this actually might mean. *General relativity* also establishes the mathematical basis for *black holes, wormholes* and other strange phenomena that today's cosmologists are excitedly exploring. **Most importantly, this physics demonstrates that many of the very things that we observe and use every day in our lives, including all things that involve gravity and time, are a direct result of the geometry of a universe that is formed and held within a type of space that requires more than our normally perceived three-dimensions.**

For more than one hundred years, we have known that our universe is constructed from more than the familiar three dimensions that were first described by Euclid. We have been learning that *gravity* and "time" are only "trickle down" shadows, artifacts or *projections* cast upon our *three-dimensional viewport* through this much larger multi-dimensional space. **A multi-dimensional geometry, which extends through regions or realms that we cannot directly see or sense, forms the fundamental space that contains our universe.**

CURVED SPACETIME

One of the more significant conceptual surprises from the mathematics of *relativity* is that lines that look straight to us are really "curved" because *spacetime* itself is *curved*. *General relativity* showed that *gravity* is how we experience the distortion or *curvature* of the *spacetime continuum* created by massive objects, such as the Sun or planets. Curved is a word we understand well in three dimensions, but in four dimensions *curved* has a different meaning, one that we cannot easily comprehend. Since we cannot directly experience or understand *spacetime*, *curved* becomes the closest three-dimensional concept to describe this new idea about the *shape* of space. Gravity appearing as a *force* is an artifact, one of many caused by the *dimensional reduction* of our *viewport*. We are not able to accurately perceive or describe such a multi-dimensional space because we evolved and have been conditioned to understand and function in just three dimensions.

We all understand that the ground under our feet is really the surface of a curved sphere, our planet Earth. However, it still feels and looks flat to our senses. In a similar way, spacetime seems "flat" when viewed in three dimensions, even though it is mathematically curved. The drawing below illustrates how a *curved spacetime* might appear in only two-dimensions.

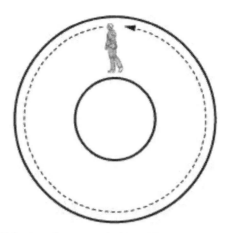

In curved spacetime, if you look into distant space you will eventually see the back of your own head. This is a two-dimensional diagram. Curved spacetime requires at least four dimensions.

Unfortunately, this way of visualizing a curved *spacetime* is misleading because it only represents two dimensions instead of four; we are only seeing the shadow of a shadow – four dimensions reduced to two. Despite this inherent inaccuracy, many physics teachers still use this diagram to introduce their students to the idea of a curved *spacetime*.

Because human brains are designed to work in three dimensions, we can use two- and three-dimensional models as tools to help us "sense" this multi-dimensional geometry. Like *Zen koans*, these models do not describe the actual geometry, but rather they function as useful tools to point us towards a deeper level of hidden truth.

MATHEMATICS AS A GUIDING LIGHT

While impossible to visualize, we can still describe and manipulate multi-dimensional geometry with relative ease using mathematics. Mathematics is a language that transcends our ability to visualize and even our need to logically "understand." The reader might now ask, "Is it fair or reasonable to be equating mathematics to things that have real meaning in our world?" History demonstrates that the answer to this question is affirmative; time and time again, our mathematical predictions turn out to have real-world implications. Our history demonstrates that mathematics has been a clear and accurate predictor and a faithful guide. ***Mathematics has served us by providing an extremely powerful searchlight to help illuminate new paths through the unknown.***

Mathematics is a language. Since at least the days of ancient Greece, there has been a lively philosophical debate about whether mathematics exists in nature and we discovered it, or whether it is our own creation. For reasons to be discussed later, at the deepest levels of creation, we discover that these two views are really one and the same. Regardless of its origin, mathematics has always provided us with glimpses into the unknown, which have completely changed our understanding of the universe. Like microscopes and telescopes, mathematics is a tool that helps us to see beyond the normal limits of our senses.

Of course, any attempts to provide human meaning to these mathematical results are completely subject to the limits of our *viewport* and current cultural paradigm. As a result, potential meanings will also shift and change as we reach new levels of individual, cultural, and scientific understanding. The more we learn, the clearer it becomes that there are vast parts of our universe that we simply do not understand. Mathematical tools provide us with theoretical clues and vistas into these otherwise invisible worlds. ***It was mathematics, alone, that gave us our first glimpse of extra-dimensional space about 150 years ago. This expanded view of space is now the key to this entirely new way of understanding our personal relationship with our universe. Mathematics can lead us to things and places that otherwise are beyond our imagination.***

VISUALIZING A MULTIDIMENSIONAL GOEMETRY

In the next section, we will begin our journey into the world of *quantum physics*. *Relativity* required us to expand our dimensional paradigm one more step: from three dimensions to four. However, as we attempt to integrate *quantum physics* and *relativity*, even this four-dimensional coordinate system becomes woefully inadequate.

About the time of Einstein's birth, several mathematicians introduced the idea that the geometry of the universe required one additional dimension, making it four-dimensional. Today, most physicists realize that a four-dimensional *spacetime* is the absolute minimum; many think creation requires even more dimensions. A significant number of physicists who have been exploring *string theory*, *super-string theory* and *M-theory* are even predicting that as many as 10 or 11 dimensions might be involved.

If we seek to understand the real mechanics and structure of our existence, we must come to the place where we accept and recognize that creation is built from more than three dimensions, even though we do not have the direct ability to perceive or visualize them. These extra dimensions are not just for mathematicians, for they are required to form and maintain the familiar structure of this physical world that we all move through every day; these invisible dimensions shape our entire experience. ***The reader's willingness to accept that our universe is built upon a multi-dimensional geometry is a prerequisite for many of the ideas discussed in this book. This hidden geometry is infinitely expansive, but it also simultaneously connects everything in ways we cannot possibly imagine. It is this architecture that makes the "impossible" very real!***

What does this multi-dimensional geometry look like? While it remains impossible to visualize from our conceptual space, we can use a three-dimensional model again to help. The nested model below is one three-dimensional way to think about additional levels or dimensions.

Eleven nested dimensions all acting as one single system. This diagram is drawn on a flat (2D) piece of paper so there appears to be an "inner" and "outer" dimension. This effect is only an illusion experienced in our limited 3D realm. In 11-dimensional space, our 3D concepts, such as "in" or "out," will have very different meanings.

We know that all the additional dimensions must fit and work together in a fully interactive way. In fully interactive multi-dimensional space, all dimensions are of equal and critical importance to the maintenance of the entire architecture; every level is equally important because if one is lost, the entire structure collapses. Like the foundation, walls, roof, and electrical system of a house, all of the parts work together and depend on each other. All dimensions also can be understood as simultaneously "inside" and "outside" of each other, and there is no dimension that can be described as "higher" or "lower." Our ideas about "inside or outside" and "higher or lower" are only artifacts created by our *viewport.* ***Creation is entirely interactively interconnected, so that all dimensions contribute equally to the whole. To speak of "higher" or "lower" dimensions as more or less important is equivalent to thinking that our hands are more important than our feet because they sit higher on our body.***

QUANTUM WEIRDNESS

Relativity was only the beginning of our journey into the strange world of modern physics. While *relativity* might sometimes seem illogical when viewed from our old "common sense" perspective, we now discover that *quantum physics* takes this type of strangeness to a completely new level.

BOHR ATOM MODEL

In junior-high chemistry classes, my generation was taught that the *atom* was the *elemental* building block of all *matter*. We were told to visualize these atoms as having a solid center, called the *nucleus,* with *electrons* circling around, creating a form similar to our *solar system.* This was a simple model and it seemed logical for the *atom* to be constructed like our *solar system* since nature tends to repeat itself. *Families* of *atoms* were identified, and we were taught that most of the differences between these *families* were due to the number of *electrons* in the outer rings of the elements. Interactions between atoms were determined by every *atom's* strong "desire" to have its outer ring filled with its ideal number of *electrons*; this allowed the *element* or *compound* to exist in a more stable, *lower-energy* state. To someone like myself, who thinks in pictures, this image, called the *Bohr Atom,* was clear, simple, and satisfying.

Unfortunately, this model of the *atom* turned out to be crude, approximate, and even quite wrong. It was the 1960s and very few high school physics and chemistry teachers understood the rapid changes that were revolutionizing their fields. By the 1920s, leading physicists began to understand that this once-convenient model had several serious, even fatal, problems. It still worked to explain many interactions between *atoms,* but it could no longer be recognized as an accurate physical representation. Niels Bohr was the central figure in the development of this brand new physics, named *quantum mechanics*, which clearly proved that the old classical model was obsolete.

Ernest Rutherford had earlier developed the simple model of the atom that I was taught; but later, Bohr modified it to include aspects of the new and rapidly developing *quantum mechanics*. This new, modified model was named the *Rutherford/Bohr atom,* but ironically – maybe because his name was easier to remember – Bohr's name became forever tied to the older model that completely pre-dated his paradigm-changing contributions.

ELECTRONS AND CLOUDS OF PROBABILITY

One of the biggest misconceptions from the old *Bohr atom* model is about the character and quality of the *electrons* themselves. As physicists learned more about these extremely small but critical components, they began to realize that *electrons* are not actual "things" spinning around a core center, like the planets around our Sun. They actually are not "things" at all, rather they are more like hazy "clouds of possibilities;" and they only "exist" in a strange *indeterminate state* with a *potential* for physical expression. This *potential* for physical expression was described mathematically by Schrödinger's *probability wave function,* the most famous equation in *quantum physics*.

For almost a century, experiment after experiment has produced data that are most easily understood by accepting the strange idea that *electrons and other small particles* do not even physically exist until the moment that an *observer* becomes involved. The act of *observation*, or awareness, is what causes the *probability wave* to *collapse* into an actual physical thing. **Electron, photons, nutrinos, and other small particles do not exist at all in our physical realm until someone or something is observing them. Only with this act of conscious awareness does one of the many possibilities for their expression become real and physical.**

These tiny (but not fully existing) "pre-things" behave like potential ideas that have yet to form. **At the same time, these are the same particles that make up all matter, including ourselves.** As bizarre as this conclusion sounds, this is only the beginning of strange and weird phenomena that we will continue to encounter as we venture deeper into the *quantum* world.

Electrons are largely responsible for all chemical reactions in nature, most of which are the very processes that allow us to function. **The bigger things in this world, such as our bodies, are regulated and controlled by the behavior of these tiny particles, like electrons, which essentially do not even exist, are extremely elusive, and always exhibit strange behavior. Since we are built from these**

mysterious components, we also are not what we seem to be. We also are mysterious and have the potential to be many things.

As Niels Bohr was very fond of saying, *"Anyone not shocked by quantum mechanics has not yet understood it."*

THE ELECTROMAGNETIC SPECTRUM

Visible Light comprises only a very small part of a type of *energy* that we call *electromagnetic radiation*. Different frequencies have been isolated and named, but together they all describe the *electromagnetic spectrum*. When we speak about *light, we are* commonly referring to a very small and limited part of this *spectrum*—the part that is visible through our eyes. *Visible light* is only those *wavelengths* that fall between "heat producing, *near infrared*" on the *low frequency, long wavelength* end of the spectrum, to the much more energetic *"near ultraviolet"* on the higher end. Above and below this narrow range that our eyes can see is an extensive continuum of many other higher and lower *frequencies* of *radiation.* Even though we cannot see beyond the range of *Visible Light*, many other species on Earth naturally see, sense, and use some of these other frequencies. However, while these *frequencies* may be invisible to our eyes, they still are extremely important to our lives.

At one end of the *electromagnetic spectrum,* we find the lowest *energy,* longest waves, and lowest *frequencies,* which we call *Radio Waves.* As the name implies, we use some of these *frequencies* as carriers for radio broadcasting. As the frequencies get higher and wavelengths correspondingly shorter, we have, in order: *Microwaves, Infrared, Visible Light, Ultraviolet, X-rays,* and *Gamma Rays.* As we move through this list, the size of the energies involved also increase dramatically. We understand the dangers of these higher *energy* forms of *radiation,* so we try to avoid exposure to high *energy Gamma Rays, X-rays,* and even *ultraviolet,* yet we constantly bathe ourselves in a sea of human-produced *Radio Waves* and even *Microwaves* from Wi-Fi, body scanners and cell phones. Some of these lower *frequencies* may also turn out to be less than completely safe—we might remember that *x-rays* were once thought to be harmless. As a general precaution, we should probably try to limit our exposure to these self-generated low-level *microwave frequencies* until we understand more about their long-term effects. The Earth's magnetic field and atmosphere naturally shield us from dangerous outside *electromagnetic radiation.*

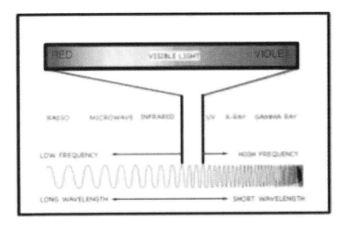

Visible light is only a small part of a much bigger range of electromagnetic radiation that we depend on for our technology and our very lives.

Without *visible light,* we would have no sense of sight, yet other species can utilize different parts of this *spectrum* for "sight." While *visible light* is responsible for much of *photosynthesis,* the critical beginning of our food chain, plants also use parts of the *electromagnetic spectrum* that are invisible to us. Some of the critical *frequencies* that plants require fall in the high *ultraviolet,* which is outside our human range of vision. ***Photosynthesis is another powerful reminder that we are all built from energy.***

Electromagnetic radiation is critical for physical life; we completely rely on it for many essential processes. Without this *light energy* from the Sun, there would be no life on Earth, at least as we understand it. Most of Earth's *heat* comes from the invisible *infrared* bands of *radiation* at the lowest end of the *visible spectrum,* but even *visible light* contributes some of the heat that fuels our planet. None of the plants and animals living on our planet could exist without this *radiation*; even those that do not require it directly, like deep sea or cave creatures, need it indirectly.

We can apply most of this book's discussions about *visible light* to the rest of the *electromagnetic spectrum*. The entire *spectrum* plays a critical role in our everyday lives, even if we are not consciously aware of this fact. It is responsible for transmitting and communicating *energy* and *information* at the speed of light throughout our planet and universe. Most of the time, when I refer to *light* I am actually speaking about the entire *electromagnetic spectrum*.

QUANTA AND THE PHOTOELECTRIC EFFECT

Today we know that when *light* strikes a solar panel, it produces electricity in a real-world demonstration of what Einstein called the *photoelectric effect*. This *photoelectric effect* occurs because "packets" of *light*, which he named *quanta,* can act like particles that strike *electrons* within the material and send these *electrons* into motion. Visualize this by thinking about what happens when the cue ball strikes other balls in a game of billiards.

The first detailed mathematical description of the *electromagnetic spectrum* is attributed to James Maxwell in the mid-1800s, and his groundbreaking physics paved the way for Einstein. Over the next fifty years, Faraday, Boltzmann, Plank, and several other physicists also studied the unusual nature of *electromagnetic radiation*. However, it was Einstein, in *The Photoelectric Effect*—the first of his four groundbreaking research papers published in 1905—who demonstrated that light traveled in discrete particulate packets. When *energy* (*mass*) is released from *matter*, it is always grouped in discrete bundles or quantities, which he personally named *quanta*. The name *quantum physics* refers directly to Einstein's discovery that *light* and all other forms of *electromagnetic radiation* exhibit this property.

Einstein's breakthrough paper on the *photoelectric effect,* along with his two papers that later became the *theory of relativity,* established this time as the single most productive period of his life. However, one of the greatest ironies in the history of science is that the name *quantum physics* came directly from Einstein's research: he was not a *quantum physicist.* He spent much of his later life fighting the extended implications of his own discovery and never fully accepted *quantum physics*. His famous quotations, "God does not play dice with the universe" and "Spooky action at a distance" were expressions of his personal criticism of quantum theory.

Einstein's *photoelectric effect* paper earned him the Nobel Prize in 1921, some sixteen years after its publication. Today, we recognize that he also deserved another Nobel Prize for *relativity.* However, at that time *relativity* was such a dramatic departure from the known paradigm that the Nobel committee was not prepared to "stick their necks out" and award him their prize. The *classical physics* paradigm was rapidly dissolving and the scientific community was extremely confused as the old, reliable laws of physics were being tumbled from multiple directions. **It was during this period of rapid new discoveries that our once-certain world began to be seen as a much more uncertain world, but also one full of infinite and amazing possibilities.**

The *photoelectric effect* also proved to be the key to uncovering one of the strangest principles of *quantum physics*—the dual or "*particle-wave*" nature of *light*.

PARTICLE-WAVE DUALITY

Einstein demonstrated that *light* could behave as a beam of *particles*, which he called *photons*. This *particle nature* of light is what actually causes the *photoelectric effect,* and this collision and movement of electrons creates the electricity in our small, solar battery-chargers; numerous, common, daylight sensors; solar, outdoor night-lights; and our homes' *photovoltaic* systems. *Light* particles (*photons*) hit the *photocells* and dislodge *electrons*, causing the loose *electrons* to move (or at least move their charge) and generate electricity. The fact that *light* can behave as if it were a beam of *particles* is a fully integrated part of our scientific understanding and contemporary lives.

However, through experiments that began in the early part of the 20th century, physicists and chemists discovered that light could also behave like a wave, similar to a sound wave. **When we look at the characteristics of light experimentally, we sometimes see light acting like a stream of particles and sometimes like a wave.** After a few well-analyzed "happy accidents" in laboratories, along with many clever, well-designed experiments that all produced the same strange results, **physicists reached the understanding that light exhibits both wave and particle behavior, but never both at the same time.** At any given moment, *light* behaves as if made from either *particles* or *waves*. Physicists call this "not-yet-determined" state of multiple possible expressions *quantum superposition. The Architecture of Freedom* describes specific experiments that reveal this split personality of *light*.

As we investigate further, this split behavior gets even stranger because what determines whether light acts as waves or as particles seems to be what the observer expects or anticipates. We are the *observers*, so our expectations somehow determine how this *electromagnetic radiation* behaves or appears. Once again, we discover that the act of our conscious participation can influence or determine the physical outcome. **This is revolutionary because it means that our best science is telling us that our own expectations somehow directly influence the expression of our physical world!**

Particles constantly "jump" to different locations, but they seem to do this in a special way, without passing through the space in-between. An *electron* may jump to an entirely different *energy* state but they never appear anywhere in the middle, either physically or in the measured values. It is as if they just disappear from one place and then reappear in another. This *particle-wave* choice and the "jumping" seem to happen instantly, which also means that the transfer of this *information* appears to happen even faster than the speed of *light*.

These extremely small *particles* also do not seem to understand *time* in the same way that we do. Individual particles striking a target can behave like a coherent wave, even if vast amounts of time separate the release of these individual particles; the passage of time does not appear to change their behavior at all. Another astounding realization from experiments that involve the timed release of photons is that these small particles can also appear to be present in two places at the same *time*. (It is informative to remind ourselves that our bodies are also made from these same tiny, but very strangely behaving, particles.)

Another important *quantum* principle is the discovery that we cannot measure both the *position* and the *momentum* of subatomic particles at the same time. If we measure *position*, the very act of measurement interferes with the *particle* in a way that makes accurate measurement of *momentum* no longer possible. The reverse is also true, so measuring *momentum* means we cannot accurately locate the *particle*. This is the *uncertainty principle*, first proposed by Werner Heisenberg, which states that we will always be uncertain about either the *position* or *momentum* of *quantum particles*. Again, we are reminded that whenever we interact, we will always change the very thing that we are observing.

I briefly mentioned one of the most interesting *quantum* properties: the ability of *subatomic particles* to communicate to each other instantly over *infinite* distances. Physicists call this characteristic *entanglement* and say that the particles are acting *non-locally,* which means that their location in space does not seem to matter or have an impact on their interaction. Einstein referred to this phenomenon as *"spooky action at a distance,"* since this was one of the implications of the experimental data from *quantum physics* that did not sit comfortably with him. *Non-local interaction* is an extreme understatement for a quality that is astounding. To understand *non-local interaction* completely requires that we radically transform our entire view of the universe, and even of life itself. Timothy Ferris, a former Berkley professor and the author of *The Whole Shebang,* describes this phenomenon using one of my all-time favorite quotes about the strange *quantum* world – **"It is as if the quantum world has never heard of space—as if, in some strange way, it thinks of itself as still being at one place at one time."** These *particles* behave as if the vast distances of our three-dimensional universe are meaningless. This new *quantum* world is clearly very different from the old world we once thought we understood so well.

MODERN QUANTUM RESEARCH

Quantum physics is the physics of extremely small things such as *particles* like *photons, electrons* and *quarks.* These all are much smaller than that which we can directly "see" with the equipment or instruments available today. Since our eyes or instruments cannot directly see them, the behavior of these invisible *particles* must be explored indirectly by mapping the trails or traces that they leave on various detectors or screens. Unraveling this *quantum* world has been a primary goal in physics for almost one hundred years.

Due to the high *energies* and speeds involved in *particle physics*, the equipment required for this research has been getting larger and larger. While scientists conducted the early experiments with devices that easily fit on tabletops or in rooms, the newest piece of equipment for this type of study, the *CERN Large Hadron Collider*, an accelerator, in Switzerland, is seventeen miles in diameter. In Texas, construction began for an even larger and more powerful accelerator that was to be more than fifty miles wide. Named the *"Super-conducting Supercollider,"* its construction was halted at the halfway point in 1993 because of politics and budgetary concerns.

Using this expensive equipment, along with simpler experimental devices and modern astrometrical measurements from space, scientists have collected a substantial amount of data and knowledge about this mysterious world of the ultra-small. From the perspective of our everyday existence, these tiny "things" behave in very unexpected ways. Within the physics community, one generally agreed-upon conclusion is that **the observer's expectations somehow affect the outcome.** Another equally strange realization that emerged from this research is that before the moment of *observation*, these *particles* only exist as a *probability wave*; they are not yet physical "things." They are, instead, in some kind of indeterminate state where they have the potential to exist. Only after an act of *observation* does this "wave of potentiality" *collapse* so that the *particles* become "real." **In other words, we create our 'real' world with our expectations.**

BRANCHES OF QUANTUM THINKING

INTRODUCTION

As *quantum physics* matured, several distinct schools of thought emerged, as philosophers and physicists tried to understand and explain the unexpected and amazing experimental results.

COPENHAGEN INTERPRETATION

The original, and for many years the dominant, *quantum* explanation is called the *Copenhagen interpretation.* Niels Bohr held the position as its champion and chief interpreter until his death in 1962. His was a tightly encamped group that believed that *quantum* weirdness was best explained by accepting the fact that these small particles only exist as *waves of probability,* or *potential,* until the actual act of *observation. Superposition* is their name for this indeterminate state of *matter* that exists before actual *observation.* At its most extreme, this view implies that there is no objective reality because it is only the actual act of *observation* that nails down all particles, and, therefore, our reality. The Copenhagen camp went even further to declare that if something is not *observable,* then it does not even exist and is, therefore, not even worth talking about. ***Reality does not exist without observation.***

MANY WORLDS

In 1955, Hugh Everett proposed a very interesting, alternative explanation, called *The Many Worlds Theory*. In a single bold leap, he introduced the radical idea that every time a decision is made or a path is chosen, the "world" actually splits or divides. With this split, another new world is instantly created and, even though we consciously observe only one world at a time, all the other possible outcomes exist from that moment on. Every time we make an act of observation or a decision, our world splits again and creates new versions that contain all the possible outcomes. All these new worlds then have the same *logically consistent* past history; but now multiple new worlds exist, expressing all paths to every possible future outcome. This process looks like an ever-branching flow chart or the branches of a tree. Another way of stating Everett's theory is that ***all possible outcomes do occur in a very real and concrete sense, and every possible outcome results in an entirely new branch of history.***

Everett's original *Many Worlds* view has all possible outcomes being "created" at the moment of decision or *observation*. This means that our three-dimensional sense of "time" is, therefore, once again involved and responsible for ordering events. ***However, beyond our "time-ordered," conscious minds, all of these possibilities have always existed, exist now, and will always exist. They therefor, are not being created at the moment of observation or decision because no such moment actually exists; we only perceive this moment due to the illusion of "time" generated by our brains. Instead, we become aware of one of these "pre-existing" worlds through our attention. This is one way to understand how all possibilities, or branches for the entire Web of Infinite Possibilities, can already exist.***

BOHM'S ENFOLDED UNIVERSES

David Bohm's *implicate order,* or *enfolded universe* theory, is the third major interpretation of *quantum* experimental results. Proposed a few years after Everett's *Many Worlds* explanation, it was not until the 1970s, when experiments confirmed the fundamental cornerstone of this theory, the *nonlocal* nature of *particles*, that it started to receive deserved attention.

This interpretation of *quantum mechanics* implies a "swirled" order to the universe, where every part is in direct contact with every other part. A demonstration, using the common salad dressing of peppered oil and water, provides one of the best explanations for *enfolding* with a familiar three-dimensional example. While sitting still, the three layers are very separate; but when shaken, suddenly the oil, water and pepper touch each other everywhere.

Enfoldment involves more dimensions than three, so it allows for an expanded type of "direct contact" to occur over any distance and through dimensions; *non-locality* can exist only because of the multidimensional structure of creation. ***Because of the enfolded nature of creation, every part of the universe is always intimately connected to every other part of the universe. Nothing can ever be hidden. Everything in the universe is always fully interactive and responsive. Every thought, observation, or action is communicated throughout and fully contributes to creation. Every thought or action from everyone and everything contributes.***

Bohm, and now many others, compare this vision of the universe to a *hologram*, where every piece of *information* is stored everywhere in such a way that we can see a record of all creation within even the smallest pieces. Later in this chapter, there is a discussion of *holographic theory* and some of its amazing implications.

Some physicists have taken the idea of enfoldment a step further, theorizing that, at some deep level, the reason for this *non-local behavior* and instant interconnectivity is that **all things that appear to be separate are really only one thing!** I believe this represents a critical and fundamental truth about our existence. *The Architecture of Freedom* expands upon many possibilities that are generated by this groundbreaking idea.

STRING THEORIES

Introduction

As physicists continue their quest for *unification,* two different and currently distinct areas of focus have emerged—*string theory* and *quantum gravity theory.* The former derives from *quantum physics,* while the latter builds upon *relativity.* Of the two approaches, *string theory* pushes the old paradigm further because it requires adding at least six more dimensions to our model of the universe. *String theory* emerged from our search for the smallest *elementary* building blocks.

Elementary Building Blocks

Many ancient cultures believed that air, water, fire, and earth were the *elementals,* or the basic building blocks of *matter.* Then, beginning with the ancient Greeks, the *atom* was recognized as the primary building block of *matter.* For 2,000 years, we believed that *atoms* could not be subdivided; they were the smallest indivisible particle. In the early 20th century, we learned that the atom itself was actually constructed from *protons, neutrons,* and *electrons.* Most physicists viewed this simplification as great progress, since three small particles had replaced a list of more than one hundred larger *elements.* However, those who felt that we had finally found the smallest pieces of *matter* with these subatomic particles were soon surprised when, in the 1960s, physicists discovered *quarks* – the even-smaller building blocks of *protons, neutrons,* and *electrons.*

These new subatomic particles behaved just as predicted and strictly adhered to the rules of *quantum mechanics.* At the time *quarks* were discovered, these *quantum* rules were called the *standard model.* Unfortunately, even with all the discoveries of small and then even smaller *particles,* there still was no acceptable way to combine *relativity* and *quantum theory.* For most physicists, the continuation of this *unification* problem indicated that something was not quite right.

Early String Theories

In the 1980s, a new theory called *string theory* re-energized the *unification* movement by presenting a new interpretation of the *quantum* experimental result. This theory proposed that even *quarks* could be subdivided, and this time the smallest elemental pieces were no longer particles, rather more like tiny vibrating *strings.* According to *string theory,* everything physical manifests from these extremely small, vibrating *strings* that exist in various shapes, some even forming loops. *Strings* are always in vibration, and the precise way that they vibrate determines specifically which types of particles they manifest. Even more interesting, the math indicates that these *strings* only exist in a minimum of ten dimensions.

String theory and its spinoffs are mathematical and theoretical in nature; there presently is no experimental proof of their existence. Since these theories are based on many extra dimensions, experimental proof will likely elude us for a long time because we have yet to conceive of a way to conduct physical experiments for testing theories that require this many extra dimensions. However, even without experimental proof, the mathematics alone has created enormous excitement amongst physicists. It works relatively well, and the new forms revealed make a special kind of sense to those most familiar with this type of mathematical exploration.

The sticky problem of *gravity* not being mathematically describable from within *quantum mechanics* was addressed when *string theory* was further "modified" to allow for the existence of the *graviton*, the theoretical *particle* that obeys all *quantum* rules and helps to explain *gravity*. With the addition of this *graviton*, many physicists saw new potential for *unification* and called this sub-theory and its variant *super-string* and *super-gravity*, respectively. Before long, however, four other fully consistent *super-string* theories that described other possible arrangements were hypothesized. The problem was that they all seemed to directly compete with each other.

M-THEORY

Finally, in the mid-1990s a *string theorist* named Edward Witten, who some feel is the most brilliant living mind on the planet (and not just for his contributions to physics), determined that the five, different, originally-proposed *string theories* were simply different views of the same theory, and could be combined into one theory if one more dimension were added to *string theory*. **M-theory, which is his further refinement of string theory, therefore requires eleven dimensions instead of just ten.**

According to *M-theory,* the source of all creation seems to be a *membrane* functioning like a multi-dimensional drumhead. The *strings* from ten-dimensional *string theories* are built from one less dimension than these eleven-dimensional membranes, and can be thought of as dimensionally-reduced slices of this *membrane*—a relationship that is similar to the sphere being seen as a circle when dimensions are reduced by one in *Flatland*.

According to *M-theory*, these multi-dimensional vibrating *membranes (also called M-branes,* or just *branes)* are the source of all material creation. In our three-dimensional *viewport*, all that we can ever see or experience of these *membranes* are the final *projections* that eventually manifest as objects in our reality. ***The particles that we interact with in our three-dimensional universe are only the dimensionally reduced artifacts, shadows, or projections that ultimately reach us from this unfathomable eleven-dimensional symphony.***

One of the most exciting potentials of *M-theory* is that it provides a pathway to combine *relativity* and *quantum mechanics* in a mathematically elegant way; this quality alone has attracted the attention of many in the physics community. While *M-theory* does explain more of the mystery and connect *relativity* and *quantum mechanics*, it continues to bother many physicists because it is completely untestable and will remain so until we find a method for experimenting with and peering into extra-dimensional space.

These *strings or membranes* vibrate, and as they vibrate they create music-like *resonances* and harmonies. These *vibrational* symphonies then pass through eight layers of dimensions (from eleven dimensions to three) to manifest as the small physical *particles* that are the building blocks of all *matter* in our realm. ***According to the various string theories and M-theory, all matter begins with vibration. In the beginning there is a "song," which is originally performed in an eleven-dimensional concert hall! That with which we interact directly is only that part of the symphony that reaches and translates into our three-dimensional world.***

UNIFIED FIELD THEORY

As discussed, the most dominant and pressing problem within modern physics has been the inability to find a way to combine *quantum theory* and *relativity*. Between these two types of physics, we have useful descriptions of all four of the fundamental *forces* that we currently have defined: *gravity, electromagnetic, weak nuclear,* and *strong nuclear*. The first of these, *gravity,* is explained only by *relativity,* while the other three are described quite well through *quantum theory*. Most physicists agree that a complete description of our universe must include all four of these *forces,* but although many physicists have devoted their lifetimes to this project, all of their attempts to combine these two theories have been repeatedly thwarted. This *Unified Field Theory* has been the Holy Grail of physics, but it was not until the 1970s, with the discovery of *String Theory*, that anyone saw real possibilities for *unification*. Another recent theory, *quantum gravity,* which derives from *relativity*—meaning that it is limited to four-dimensional *spacetime*—also holds some promise for *unification*.

When we include eleven dimensions into our calculations, as *M-theory* requires, suddenly *quantum mechanics* and *relativity* start to fit together quite naturally. Calculations become mathematically elegant because the equations balance and the pesky problem areas begin to self-resolve. With these extra dimensions in the mix, *quantum physics* and *relativity* start to make mathematical sense together. *M-theory* may be the best current trailhead for leading us to the long sought-after *unified field theory*.

WHAT WE SEE

Clearly, the *deterministic,* machine-like world of *classical physics* is no longer an accurate description of our existence; the new reality includes aspects that are known to be *probabilistic* and others that are still quite mysterious.

Our expectations originate from old patterns built into our three-dimensional consciousness. Our experience is actually only the *artifact* or *projection* that we construct from energetic *information* that would make much more sense if it could be viewed and conceived through an expanded *viewport* containing more dimensions. We only witness the "trickle-down" shadows from this deeper reality, and this means that the three-dimensional world we experience is only an illusion.

Of course, this illusion also includes all of our attempted explanations for these experimental results. For example, the fact that we are surprised when *electrons* or photons do not behave differently when their rate of release is slowed down is only because we always see our world through the "lens of time." If we clearly understood that "time" is just an organizing concept for our brains, then we might, instead, expect these results.

In a *multi-dimensional* world, where "time" does not divide events and an *infinite* number of multiple universes are interacting in ways we can't now imagine, speed and distance have very different meanings. We just cannot think of the universe in our normal, limited, three-dimensional terms and expect to be able to understand these *quantum* experimental results. Again, we are only seeing the shadows or images and thinking that they are real. When we try to understand or explain our multi-dimensional universe from our limited *viewport*, we will always fall far short.

One way to gain a broader level of "understanding" is through learning to relax our habitual need to explain all things from within our logical and time-ordered system of thought; we can practice being present as we sit, experience, and just be a part of this great mystery: "become a passerby." This is a very effective form of meditation and along with relaxation, a deeper type of understanding gradually forms. *The Architecture of Freedom* includes a much more detailed discussion of this idea.

Through our new awareness built upon this physics, we can start to imagine a different type of universe that might exist outside of time. *In this universe, all worlds lie together within a unified, multi-dimensional soup, where everything is so deeply enfolded that all possible outcomes are instantly adjacent and available, from any place, at any time. Any piece of information from anywhere or anytime is directly and immediately available to any and every other point in this web-like system. This Web also acts like a holographic storage system; no matter where we look or how small the part, all the information about every possibility is always locally and immediately available.*

As we discuss these or any interpretations, it is necessary to remember that ideas in *quantum physics* and *relativity* are quickly evolving, and many things will change as our new paradigm evolves. *These theories are discussed with the ever-present caveat that any explanation created from our three-dimensional conceptual mindset and language will always miss the mark because we are exploring something that is far beyond our current ability to discuss or understand. This is the nature of our deepest and most beautiful journey.*

COSMOLOGY

Cosmology, the study of the origin and development of our universe, involves all of the branches of science. Modern *cosmology* began with the paradigm-shifting work of Kepler, Copernicus, Brahe, and Newton, but the last 100 years of *relativity* and *quantum physics* has revolutionized the field. Just as *classical physics* helped usher us out of the "flat-Earth" paradigm, *relativity* and *quantum physics* have stretched the limits of our imagination by introducing many brand new ideas about the potential nature and *shape* of our universe. **Since the biggest things are made from the smallest, we are now discovering that there is far more connection between "out there" and "in here" than we ever imagined.** Because of these discoveries, *Cosmology* now includes dramatic new approaches to issues such as *gravity*, empty space, *time*, consciousness, and the *observer*. In addition, a steady stream of new discoveries uncovering mysterious things such as *black holes* and *dark matter* are rapidly and completely changing our old ideas about the nature of our universe.

DARKNESS WITHIN OUR UNIVERSE—DARK MATTER

One of the most unexpected discoveries in modern *cosmology* occurred when new technologies and methods allowed scientists to more accurately analyze the relative motion of the various objects of our known universe. They discovered that there simply was not enough visible *mass* present to produce the relative motions of the stars and *galaxies*, account for rotational speeds of *galaxies,* and explain the *gravitational bending* of light from distant stars. The amount of *gravitational matter* necessary to explain these inner workings of the universe was not just a little bit off. It was not even off by double—it was off by at least a factor of five! This means that the visible *matter* is less than one-fifth of what is required to produce the relative motion of everything we can see or measure. By some calculations, we only see and understand one-tenth of the *gravitational material* that must exist.

One of two things is probably occurring: either all physicists are making a colossal mistake in theory or calculations, or between eighty and ninety percent of the *gravitational* "stuff" that makes up our known universe is completely invisible to our senses and technology. We have no idea what this invisible *gravitational material* is, but we know that it affects and interacts with everything else that we can see. Since this mysterious "stuff" emits no light and we cannot see it directly, we call it *dark matter*—meaning it is "dark" to our tools and senses. We now include *dark matter* in all contemporary calculations about the cosmos, even though its existence only became fully recognized in the 1990s.

In addition, scientists have recently discovered that *dark matter* is not the only hidden "dark" stuff in our universe. There is also *dark energy,* of which there is even more than *dark matter.* Before we examine *dark energy* in more detail, we first need to understand what is being described when cosmologists talk about the *shape* of our universe.

SIZE AND ARRANGEMENT OF UNIVERSE

Looking at the extents of our known physical universe and wondering about how big it really is, scientists imagine that there are five basic possibilities. The first possibility is that the universe could be *finite*, alone in creation, and have an edge somewhere beyond the limits of our present view. The second possibility is it could be alone, but be *infinite* and extend forever. The third possibility is an interesting hybrid of the first two: the universe could be *finite* and *closed* in a bubble-like fashion, but it might be only one of an *infinite* number of these *finite* "bubble" universes. The fourth possible combination is that it is *finite* and one of a *finite* number of other universes; and the fifth combination is that the universe is *infinite* and one of an *infinite* number of other *infinite* universes.

The first possible configuration, the lone, *finite* universe, is the easiest for most of us to understand. With this configuration, the universe would simply end somewhere beyond our *cosmic horizon,* which today is a distance of about 40 billion *light-years* in any direction.

The light from the most distant stars that we can see today originally left those stars about 15 billion years ago, when these stars were much closer to us. Since then, the universe has been rapidly expanding, so our visible universe has now expanded in size from 15 billion *light-years* to about 40 billion *light-years* in every direction. Since this particular configuration is *finite*, somewhere beyond the limits of our vision the physical universe must somehow end. This sounds very similar to the way our "flat-Earth" ancestors imagined their flat pancake world to have an outer edge. (Am I revealing a personal prejudice here?) If this particular arrangement was the full extent of our universe, it would be *finite,* but it would still be full of amazing possibilities. Today, we have observed hundreds of billions of *galaxies* and each one contains hundreds of billions of stars. Even without additional regions or dimensions, this type of universe would still be large enough for many other worlds that might be similar to our own.

To understand the other four possible arrangements, we will first need to become familiar with a few new terms and concepts. When they speak of the *shape* of the universe, they are referring to a specific mathematical quality that lies beyond our three-dimensional ability to visualize.

We understand from Einstein's *relativity* that a large *mass*, such as a star, deforms four-dimensional *spacetime* locally, near the large *mass,* and we call this deformation *gravity.* All the *matter* and *energy* in the universe, because they are two forms of same thing, also work *non-locally* and in unison to deform the entire universe in some particular way. We call this larger deformation of the universe its *shape.* To simplify our discussion, we will use well-understood, three-dimensional forms such as ball, pretzel, saddle, and tabletop to describe these shapes, while always remaining aware that these are only approximations of the actual four-dimensional forms.

These three-dimensional forms are only approximations and, therefore, different from the actual *shapes* that would exist in a *multi-dimensional* universe. When we speak about the *shape* of our universe, we are referring to its shape within *spacetime,* a realm we cannot see or really even imagine. Since this realm is built from more than our familiar three dimensions, we have no spatial concepts that allow us to describe or visualize *spacetime.* These are *shapes* that lie outside our three-dimensional *viewport* and our *conceptual horizon.*

Physicists describe three potential *shapes* for the universe: it could have a *positive curve*; it could have a *negative curve;* or it could be *flat* with a zero *curve.* There are other variants and possibilities, some of them very specific, but these three *shapes* are the most important ones to understand for this level of discussion. The direction of *curvature* (*positive, negative,* or *flat*) is determined by the total amount of attractive *mass* and *energy* that the universe contains in a given area—

its density. A higher density (more *matter* and associated *energy)* will produce a *positive curvature* – a ball *shape* is the closest three-dimensional equivalent. A lower amount of energy and mass will produce a *negative curvature,* with a saddle *shape* being the best three-dimensional equivalent. It takes just the right amount of gravitational material to produce a zero curvature or *flat shape;* in three dimensions, a tabletop describes this shape.

Knowing its *shape* helps us understand whether our universe is *infinite* or *finite.* A *positive curvature* leads to a closed system that is always *finite.* A *negative curvature* usually, but not always, results in an *infinite* universe, while a zero *curvature (flat)* means that the universe **must** be *infinite.*

There is a possibility that the universe could *curve* back on itself and twist in some unexpected way that resembles a four-dimensional racetrack, which we will call a pretzel *shape.* Then, instead of looking deep into eternity, we might be looking at some of the same objects multiple times due to multiple loops and twists around the twisted track. This idea evolves from Einstein's work, and the "curved space donut" diagram, shown at the beginning of the discussion of *relativity,* describes the simplest version of this possibility. If the universe *curves* in this way (convex or ball), it results in a very big universe, but not an *infinite* one; this type of universe has been called a *bubble universe.* If ours turns out to be a *bubble universe,* it could be all alone, one of a *finite* number, or one of an *infinite* number of these bubble universes.

***However, a substantial amount of new, but very reliable, data is surprising many cosmologists because, if correct, it means that our universe has exactly the right amount of mass—including dark matter—to make the universe flat. <u>This also means that, to the best of our current scientific knowledge, the universe must be infinite.</u>*[2]**

It is also important to understand whether the universe will continue to *expand,* accelerate its expansion, or eventually start to *contract.* This, after all, will be a critical factor determining any "future" for our Earth! The mathematics of a *flat universe* is particularly elegant

---

[2] This new data was compiled from the WMAP, *Wilkinson Microwave Anisotropy Probe* satellite, which was launched in 2001. Its seven-year mission was to make fundamental measurements of *cosmology,* including measuring *cosmic microwave background radiation*.

because in a *flat universe* the total overall *energy* will be zero. Therefore, the *flat* universe is one form that could possibly have emerged from a net-zero *energy* condition during a creation process. Analyzed mathematically, we find that this *flat* universe, without any other input, will continue to *expand*; however, over time the *rate of expansion* will slow down.

DEEPER DARKNESS—DARK ENERGY

When Einstein presented *relativity* at the turn of the last century, he personally was convinced that the universe was stable and *non-expansive*; but to his dismay, he discovered that his own equations demonstrated otherwise. In the late 1920s, Edwin Hubble confirmed by observation that the universe was indeed *expanding*, but he could not yet tell if this expansion was slowing down and might someday reverse.

If continuing or accelerating expansion is occurring, then some powerful *force* must be pushing out: one that is more powerful than the opposing contraction caused by *mass* induced *gravity*. It is relatively easy to understand why this is so. If we throw a ball straight up in the air, at first it travels quickly, but then it gradually slows down (*de-accelerates*) until it eventually stops and falls back to Earth at an ever-increasing rate of speed. The thrown object eventually falls unless there is some additional *energy* added into the system, such as the firing of a rocket engine, to overpower the gravitational *force* acting in the opposite direction.

A single impulse event such as the *big bang* is analogous to the act of throwing a ball up into the air. If the universe began with a single *big bang* with no additional energy added later, then, due to gravity, the outward movement must eventually stop and reverse itself.

Through the years, we have learned, using Earth-bound telescopes and instruments, satellite observation, and, finally, the WMAP, that the *expansion of the universe*, while slowing down, is not slowing down as quickly as would be expected from a single impulse *big bang* kind of event. Some unseen additional *force* must be working against *gravitational mass attraction* and be responsible for the continuing expansion of the universe; without this extra *force* pushing out, the universe would eventually start to contract. From the best current data, most cosmologists are quite sure this contraction is not going to occur, so some repulsive *force,* opposing *gravity,* must be adding to and reinforcing, the initial *big bang.* **This means that, in addition to dark matter, there is also a mysterious, repulsive energy that we cannot see or measure directly.**

This *energy* is now fully recognized and named *dark energy.* Like *dark matter,* it is not visible to us, but we need it to explain the types of movement we are observing in our universe.

Even though *matter* and *energy* are equivalent, *dark matter,* which makes up at least 80 percent of the *gravitationally attractive mass* of the universe, and *dark energy* are very different things. Our naming of this newest unknown influence was somewhat unfortunate and inconsistent. *Dark energy* is not just another form of *energy* because normal *energy* is convertible to *mass* and is, therefore, attractive or *gravitational* in nature. *Dark matter* produces an *attractive "force"* so it behaves just like other matter: it *gravitationally attracts.* Dark *energy* counteracts the normal *gravitational* action of *matter* and *energy*; it results in a *repulsive "force,"* which makes it completely different from normal *energy.* **We have absolutely no idea what this outward "force" is, how it works, or where it comes from. However, we have calculated that it represents 73 to 76 percent of all the energy and matter in the universe.**

We have only named this *force "dark energy"* because, again, we simply cannot see it. *Dark matter* and *dark energy,* while very different, do have three things in common. First, they both affect the size, shape, and expansion of the universe in significant ways; second, we can't "see" them; and third, we know almost nothing about them.

Let us take a closer look at the astounding breakdown of the composition of our "known" universe. **The most recent WMAP III**

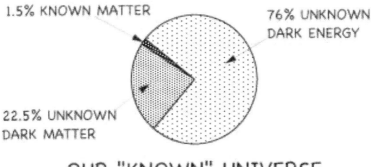

OUR "KNOWN" UNIVERSE

analysis indicates that the universe is 76 percent dark energy, 22.5 percent dark matter, and only 1.5 percent baryonic matter. (Baryonic matter is normal physical matter such as cosmic dust, planets, stars, asteroids, etc.).

When we examine more deeply that very small percentage of *baryonic matter* that is found in our universe, we find that even the very atoms which make up this small part of everything that we think we "understand" are composed mostly of empty space. This empty space within the atoms is also filled with *dark energy. **The actual amount of physical "stuff" that we (and all the stars, planets, asteroids and cosmic dust) are made of is, at the most, a miniscule proportion of the observable parts of creation. We know little or nothing about the vast majority of what gives form to our universe, and therefore form to our bodies! Everything we think of as matter is really energy, and the majority of the energy shaping our universe is completely mysterious and invisible to our senses and our devices. In addition, when we observe our universe, we are only looking at a tiny part of creation because we have not yet addressed any of it that is farther away than 15 billion light-years, smaller than a quark, or existing beyond three dimensions.***

Over the last one hundred years, we have made enormous advances in our understanding, but we have also learned that the deeper we look into the nature of the cosmos, the more we uncover a rapidly growing mystery. With each deeper probe, the number of new discoveries which cannot be rationally explained seem to be expanding to occupy a much more significant part of our awareness. The deeper we explore, the more we encounter this ever-expanding vastness, where we continue to discover new aspects, components, and perspectives that exist beyond our comprehension. ***The more we learn, the more the mystery appears to grow.***

This discussion has been focusing on just the visible or measurable three-dimensional portion of our universe. Even if our universe were only three dimensional, it appears that in some form the universe has room for a never-ending vastness: a wondrous and *infinite* expression of creation. ***If our universe is shaped this way, then we can certainly say that it is likely that other "worlds," just like ours, exist within this never-ending vastness. If, however, the universe is also infinite, there can be no real debate among physicists and mathematicians about the likelihood of other worlds exactly like***

ours. If it is infinite and also outside of time, then this very strange and unbelievable outcome is an absolute given.

Why this is so all hinges on the meaning of the concept of *infinity*. Since there is likely to be confusion about the mathematical concept of *infinity*, I will discuss and explain it in more detail before we proceed.

UNDERSTANDING INFINITY

"If the doors of perception were cleansed, everything would appear to man as it is, infinite."

William Blake, "The Marriage of Heaven and Hell".

Our conceptual minds require relationships to function; we are constantly referencing or relating new experiences and ideas to older concepts that we already know and understand. Our minds tend to work more easily within a certain range of numbers and sizes. Easily understood are numbers like one, ten, one thousand, or even fractions, such as one-quarter. A bit more difficult to understand, at least until we have some practice, are quantities such as one million dollars (six zeros), the Earth's eight billion inhabitants (nine zeros), 18 trillion dollars of U.S. debt (twelve zeros), or, on the other end of the scale, very small numbers, such as the width of a human hair (100 microns or 4/1000 of an inch). While these numbers may be quite large or small, they still directly relate to our lives. However, when we start looking at distances between ends of the visible universe (the enormous) or size of the particles that make up *matter* (the minuscule), our ability to comfortably understand these numbers falls short because we have nothing in our direct experience to relate to them. We can give these quantities creative names such as *google* which is a "one" followed by a hundred zeros, or a *googolplex*, which is a "one" followed by a *google* of zeros; but we still cannot really understand their size. Even those who work with these numbers on a daily basis often cannot fathom what numbers of this scale actually mean.

Any fixed number, minuscule or enormous, does not represent infinity. Infinity cannot be described with a number—it is a concept that lies beyond a number. If we can assign a number or amount to something, then it is not infinity.

Infinity is a mathematical idea that everyone can easily discuss and visualize because it is a concept that we can understand fully from within our three-dimensional realm. Actually, we only need one dimension to explain *infinity*.

If we pick a number and add one to it, and then take that new number and again add one to it, we have the common process called counting, which does not ever need to end. We will never reach a last and final number, no matter how many times we repeat this process. This is *infinity*. If at any time we stop the process of counting, no matter how large the number, we are no longer describing *infinity*.

Another way to imagine *infinity* is to pick a line of any length and divide it in two. Take one of the pieces, divide it, and keep repeating the process. At any point the line may look extremely short, but there will always be a midpoint to every line, no matter how short the line. This is also an *infinite* process because it will go on and on, forever. Actually, any line can be divided an *infinite* number of times and, therefore, any line contains an *infinite* number of points. *Infinity* can be fully understood using only one dimension—a line.

An *infinite* universe means that no matter how many atoms, cells, organisms, planets, solar systems, galaxies, or universes we have encountered, there will always be more to be found; and our discovery of new aspects or regions will continue and never end.

OUR UNIVERSE IS INFINITE

I have briefly outlined several lines of reasoning that all lead to the conclusion that we live in an *infinite* universe. *So far, we have arrived at this conclusion using only three-dimensional concepts. If we add additional dimensions to this structure, any remote chance that our universe is "finite and alone" will quickly disappear because, with these added dimensions, the additional pathways to the infinite become endless.*

While a finite universe based on fixed and limited bounds might be more easily understood and therefore comfortable for us, it has become impossible to avoid the pressing weight of evidence demonstrating that we live in an infinite universe. The only real questions now seem to be: in what ways is the universe infinite and exactly what does an infinite universe mean to our lives?

94

OTHER WORLDS EXACTLY LIKE OURS

Even though the universe as a whole is *infinite*, our Earth is still *finite*. There are an enormous number of *atoms* that make up our Earth, but that number can still be determined. A calculation, based on the *mass* of the Earth, estimates the number of atoms that make our planet to be near 10^{50} (or the square root of Google). This, of course, is a very big number, but because it is an actual number, it is still *finite*. All these *atoms* can be moved around and rearranged in many different ways, but eventually this *finite* number of atoms will run out of new possible arrangements; **the number of possible arrangements of the Earth's atoms, while enormous, is still finite.** This is not a theory or speculation—this is a completely rational and mathematical fact.

Imagine a simple example. Let us say we have three shirts, two pairs of shoes and one pair of pants, and you want to try the total number of combinations. After a little exploration, we will discover that this number is three (number of shirts), multiplied by two (number of shoes), multiplied by one (number of pants), or six total different possible combinations ($3 \times 2 \times 1 = 6$). Test this at home. You only can arrange a fixed number of things in a certain, fixed number of ways; the number of particles that make up Earth (10^{50}) is fixed, so they can only be arranged in a *finite* or limited number of arrangements ($10^{50} \times 10^{49} \times 10^{48}$, etc.). *In other words, a finite number of things can only be arranged a finite number of ways.*

In an *infinite* universe, these same types of *particles* or *atoms* that are found on Earth will reappear somewhere else, again and again, forever. There will actually be an *infinite* number of every type of *particle,* and they will collide, interact, and arrange themselves in different groupings an *infinite* number of times. *In an infinite universe, eventually all the possible combinations for arranging any fixed collection of particles will have been tried, so somewhere in this infinite universe these arrangements must start to repeat. This will happen over and over; and eventually, somewhere, an exact copy of Earth will be formed. Then, as amazing and impossible as this seems, because the universe is infinite, this exact duplication of Earth will reoccur again and again.*

It is a mathematical and logical fact that an infinite number of exact copies of Earth will exist in an infinite universe. This is a very difficult but critical concept, so I will state this in a slightly different way. *In an infinite universe, every one of the finite particles making up Earth will, by mathematical necessity, appear again and again, somewhere, over and over, forever. These particles will combine an infinite number of times in every possible arrangement; there is no end to the number of times these particles can combine into different combinations. Since the Earth only contains a finite number of particles, every so often the exact number and type of atoms that make up our world will be repeated, and the repetitions of the Earth's finite set of atoms will then occur in every possible arrangement. Sooner or later they will be arranged so that they exactly duplicate our Earth, along with everyone and everything on it! Then, this will process will repeat itself again and again.*

This means that somewhere, within the *infinite* expanse of our universe, every atom on this Earth will be exactly duplicated and connected to all other atoms, in exactly the same arrangement as on Earth. *These exact duplications will contain every particle and atom, arranged the same way as our world, duplicating everything in it. In an infinite universe, this process goes on forever, so there will actually be an infinite number of these exact copies of Earth!*

If the universe is just enormous, this may or may not happen. If the universe is infinite, then it has to happen.

Most of us, when confronted by the awesome nature of this proposition, will logically reason that it is ridiculous or that the odds of that happening are so small that it will never really happen! To our "dimensionally-bound" conceptual minds, the conclusion that there exists an infinite number of worlds just like ours sounds like nothing more than raw material for science fiction. It is useful to remember that the "flat-Earthers" once believed the same to be true of a spherical Earth. *However, this is how our universe is actually built; and once we break through our normal (but inherited and cultural) constraints and begin to see our universe for what it really is (full of infinite possibilities), then everything that we think we understand about life will instantly and forever change.*

EXTRAPOLATING INFINITY INTO EXTRA DIMENSIONS

The idea of *infinity* survives completely intact when it is expanded from one to two dimensions, and then again, from two to three dimensions. It therefore seems likely that expanding the idea of *infinity* to four dimensions or more will not significantly restrict or reduce the meaning of *infinity*. When we extrapolate any concept into extra-dimensional space, it is generally understood that the number of new possibilities will in fact increase. The concept of *infinity* already contains never-ending possibilities, so it easily survives such a transition intact; *infinity* expanded into multiple dimensions is still *infinity*.

Even if we had determined that our three-dimensional universe was *finite*, any conclusion that the rest of creation is also *finite* would quickly disintegrate when we expand our *reference frame* to include extra dimensions. These unseen dimensions can be thought of as simultaneously hidden "beyond" and deeply *enfolded* "within" our three-dimensional universe. Because an extra-dimensional perspective introduces many additional ways to imagine the *infinitely* vast "size" of our universe, it creates a lot of "room" for new types of possibilities. These extra dimensions also provide important mechanisms for explaining the deep, instantaneous connections between all the physically distant parts. This instantaneous interconnection throughout our universe, or *non-local* behavior, is something that simply cannot be explained using three-dimensions alone.

We can actually describe the universe as *infinite* in two different directions: "going out" through never-ending three-dimensional space, and "going in" through the unseen dimensions. *Infinity* is always being expressed at multiple levels throughout creation. As we continue to uncover new paths that illuminate the *infinite* nature of our universe, we are also beginning to discover just what this might actually mean for our lives.

Much of this book is devoted to helping the reader understand and become comfortable with this conclusion about the *infinite* nature of existence. Some readers will resist this leap because, even though this arrangement is extremely likely, it has not yet been scientifically proven and may remain unproven for some time. ***However, also unproven is the opposite conclusion, that the universe is finite. Finite systems are much easier to understand, quantify, and describe than infinite systems; therefore, it is typically much easier to prove that a system is finite. The fact that the universe has not yet been proven to be finite is, by itself, very revealing. This single***

realization throws tremendous weight towards the conclusion that our universe is infinite!

AN INFINITE MULTI-DIMENSIONAL UNIVERSE?

I am far from alone in my conclusion that our universe is *infinite* and *multi-dimensional* because many, probably most, contemporary cosmologists and physicists also believe that this is the true form of our universe. Much of what this book proposes would still be true in an *infinite,* three-dimensional universe. ***However, the addition of extra dimensions, enfolded in ways which we cannot see or really understand from our viewport, allows for additional possibilities of interconnectedness, timelessness, and unity,*** which can finally explain strange experimental results and much of the hidden mystery in our lives. The added dimensions permit a kind of interconnectedness that is not possible in a three-dimensional universe; also they allow room for an *infinite* amount of mystery and a never-ending *evolutionary* process.

Even if someone with extraordinary talent could visualize or understand four or more dimensions of space, they would still be incapable of communicating this meaning to the rest of us. Our cultural imagination reaches its limits somewhere on our side of the veil that hides four-dimensional space from our senses and consciousness. This inability to communicate about "glimpses of extra-dimensionality" was precisely what happened to the main character in *Flatland* when he tried to describe his three-dimensional experience to his two-dimensional peers. He discovered that he had no effective words or ideas to communicate his encounter, even though he fully understood everything while it was happening. He even found that once he tried to use words and concepts, his own clear understanding became muddled and confused as the limiting nature of language and concepts effectively eroded his deeper "vision."

We simply do not have any reference for what conscious life utilizing this *infinite* space would be like. Any shift in the meaning of *infinity,* as we expand it into more dimensions, must allow for even more possibilities and connections than those we might rationally deduce. *Infinity* in multiple dimensions will involve *qualitative* changes that go far beyond our present ability to imagine.

Some *cosmologists* argue that even if space is *infinite*, there is still a limit to the number of worlds that would be like our own. They argue that "evolution takes time and because of the amount of time that is required for the Earth's evolution, there are physical 'time' limits to the number and kind of duplicates of Earth, no matter what kind of spatial arrangements are possible and available." They base this argument on their perception of a real "time" constraint. Since the universe is only about 15 billion years old, they argue that there has not been enough "time" for all these other parallel Earths to have fully evolved, even if there were enough space and material for an *infinity* of them.

The weakness with their reasoning is that they base it entirely on three-dimensional logic and linear "time." If our universe was only three-dimensional and "time" was linear, then these limits would exist; but we understand that our sense of a constrained linear "time" is also only an illusion. This argument fails to allow any possibilities beyond our old three-dimensional *conceptual horizon,* and it does not even utilize our best current scientific understanding of "time."

BUBBLE UNIVERSES

If cosmologists have made significant miscalculations, the universe might close back on itself like a sphere or a pretzel. If this turned out to be our true *shape*, our universe would then be a *finite* "bubble" universe. However, even then it might be only one of an *infinite* number of other very different, similar, and identical "bubble" universes. If there are an *infinite* number of these, then we already know that, even though each universe is *finite,* there will still be *infinite* room for every possible arrangement. Within this *infinite* series of *finite* bubble universes, there would still exist an *infinite* number of universes that are identical to ours in every aspect.

THE UNIVERSE SPLITS INTO AN INFINITY OF UNIVERSES

I have just stated that there may be entire universes, not just worlds, identical to ours. ***Due to the very nature of infinity, in an infinite cosmos—an infinite number of full universes, which are exactly the same as our universe also exist.*** Again, by definition, no matter how impossible this seems, this is exactly what must happen in infinite space, both mathematically and in the real physical expression.

Every time we make a decision, the entire universe splits. Every possible variation of each decision and every possible outcome for the entire known universe also splits off and therefore exists, expressed throughout creation as an *infinite* number of identically replicated universes. For example, perhaps one leaf fell off a particular tree in one universe. Every other possible variation – two leaves falling, three leaves falling, etc. – is expressed somewhere as an entirely new universe. An *infinite* number of precise replicas of every possible variation for our universe can be found within the *infinite* extent that is creation.

As bizarre and impossible as this might seem, if our universe is infinite, then because of the very definition of infinity, this is an absolute fact. To our three-dimensionally bound minds this sounds crazy and insane, but this is not a theory! If creation is infinite and marching "time" is not a limiting factor, then there are multiple whole universes exactly like ours, along with every possible slight variation. Also, these copies are completely enfolded so that they are always in instantaneous and intimate communication with everything else in existence. It is truly an amazing Web of Infinite Possibilities.

Our specialized conceptual minds can make this idea very difficult to comprehend, or even believe. It seems too fantastic, *and it is fantastic, but not in the way we might be thinking.* Our concepts about numbers, distance, and space do not easily allow us to process and understand ideas that are truly without limits. *The real difficulty is not that space is so expansive, but that our minds and senses are not; they are designed, built, and programmed to work only within certain limits.* One day it will be only these old limits that seem strange.

In this infinite universe there is a one hundred percent probability *that there are other exact "copies" of each of us, living our lives and making many of our same decisions. In many of these universes, choices are made that are ever-so-slightly different. Every time we make any decision, thought, or movement, somewhere, other people just like us are making that same decision, thought, or movement; and at the same time others just like us are making different choices. At every moment, our universe divides so that every possible choice is fully realized and expressed somewhere in our infinite web of woven interconnectedness.*

This idea is the structural foundation of our universe and core principle of *The Architecture of Freedom,* so it will be restated many ways throughout this book. ***Today, many scientists and writers are calling this, our amazing infinite and intimately woven universe, the Multiverse.***

EXPANDING UNIVERSE

We have examined how our universe is structured from mostly empty space, which is shaped through *energetic* interactions, and how this same type of arrangement repeats itself when we examine smaller or larger systems. We see this pattern in our atoms and subatomic particles, and we see it in the planets of our solar system, the stars of our Milky Way galaxy, our region of *galaxies,* and the known universe with its billions of *galaxies*. All of these forms follow a similar, repeating pattern of vast amounts of what appears to be empty space and very little, if any, actual "solid" material, with everything being held in place by *energetic* interactions. Solid-appearing *matter* is, at the most, a minuscule component of our universe; it is universally recognized that our known universe is almost entirely constructed from empty space and *energy*. ***Since we also are made of atoms, we too are almost entirely empty space and energy.***

However, scientists have also recently discovered that empty space is not really empty at all. As I write, fresh insights into the nature of "empty" space continue to lead us towards an evolving understanding that this empty space is filled with something that is extremely interesting, but not at all understood.

When Einstein first realized that his theory of *relativity* predicted an expanding universe, he became greatly disturbed, which caused him to modify his theory by adding the now infamous fudge factor: the *cosmological constant*. This fudge factor allowed his math to predict the *stable* universe that better fit his sensibilities. However, less than a dozen years later, lawyer-turned-cosmologist Edwin Hubble discovered that the universe was indeed *expanding.* After Hubble's discovery, Einstein began to describe his own "after-the-fact" alteration of the original *relativity* theory as the greatest mistake of his career. It was not until much later, in the 1970s, that *cosmologists* began to really uncover a fuller picture. As then determined, not only was our universe expanding, but the average *velocity* of expansion was about three times the speed of light. In the time that it took for the light to reach us from the farthest stars—just over 14 billion *years*—the outer edge of this visible universe expanded outward in every direction by an additional 30 billion *light-years*. This means that in the last 14 billion years, the width of the known universe has more than tripled.

Wait, you might think – nothing can travel faster than light. As it turns out, the stars themselves are not moving at that incredible rate into some empty void. Instead, space itself is expanding at that rate. Things contained within space cannot move faster than light, but space itself seems to be able to do anything it wants. As mentioned above, there seems to be at least one other thing that can also move faster than light. Since non-local interaction is instantaneous, somehow information can also travel faster than the speed of light.

When we think about the stars moving away from us through expansion, we naturally imagine three-dimensional space and not *spacetime*. It is impossible for three-dimensional beings to visualize expanding space because we have no understanding of the structure or a "container" into which this space is expanding. What is it that contains this three-dimensional space and allows it to expand?

The common classroom analogy, blowing up a balloon, is not an accurate model because the balloon is only expanding into three-dimensional space, a realm that we already understand. However limited, this demonstration still is useful to help explain the principle. Blow up a balloon partially, but don't tie it. With a marker, make two or three dots on the surface of the balloon and measure the distance between the dots. Imagine that these dots represent galaxies. Think of the surface of the balloon as a two-dimensional surface, just like the surface of the Earth; both feel "flat" to any inhabitants, even though we know the surfaces are actually "curved." Now inflate the balloon fully and again measure the distance between the dots. This distance is now greater because the "flat surface" expanded as the balloon was inflated. What is this two-dimensional "flat surface" of the balloon expanding into? It is expanding into three-dimensional space. Now we can scale this analogy up one more dimension to provide a better understanding of our expanding universe. ***We only inhabit the "flat surface" of three-dimensional space, and when this surface expands, it must be expanding into four-dimensional spacetime!***

BIG BANG

The *Big Bang theory* was the first and most widely adopted cosmological theory about the creation of our universe that derived from *relativity*. The theory was first introduced in the 1930s, but we did not use the name *Big Bang* until some 20 years later. As the name implies, it proposes that our universe began as a very dense, hot, and microscopic *mass* that very rapidly expanded during a unique event that occurred 15 billion years ago. The actual rate of expansion required to fit the most current data is so dramatic and rapid that this event makes *supernovas* look puny. A more recent version of the theory introduces two phases of *Big Bang expansion*: the initial moment of the actual "bang," and then a secondary, more rapid *inflation.* Both phases are required to explain the even distribution of residual background *energy* that is spread throughout the universe.

There are several aspects about the *Big Bang theory* that do not ring true for me. The theory is built upon "time," so combining all of its implications, such as the splitting of ten-dimensional space and *inflation*, it seems to me that we are trying far too hard to make the known data fit into a three-dimensional concept. This *Big Bang* beginning does not seem to follow the normal pattern of most natural processes; nature tends to repeat, reuse, and recycle; therefore, this massive, one-directional process seems quite unnatural. A related idea proposed by some physicists is that the *Big Bang theory* caused ten-dimensional space to split into two smaller-dimensional realms, and the part we now live within is a space that contains only the four dimensions that we understand and experience as *spacetime*. The other part with six dimensions is said to have curled up into tiny pieces that are so small that we cannot see them with any of our measuring devices. Are we so constrained by our old conceptual paradigm that we must still resort to describing invisible things as "very small"?

To me, this dramatic creation story sounds strangely similar to some from the world's major religions. It is my clear sense that, again, "time" and *dimensional limitations* are interfering with our ability to comprehend what is really going on; a creation process unfolding in multiple dimensions simply cannot be explained using only our old concepts, thoughts, and language.

If we could stand back and see all of creation from a broader, multi-dimensional perspective, this *Big Bang* "event" might not look much different from the top of a cresting wave in the ocean. From this bigger picture, we might see the *Big Bang* as just another marker in the rhythmic wave of creation passing through the always-vibrating and interconnected fabric that manifests as the *Multiverse*. The *Big Bang* might look special and important from our reference point, but from a broader perspective, it describes just one part of a *vibratory wave* that is shaping our universe–a wave that has no beginning or end because it occurs outside of time. Ultimately, the *Big Bang* is just a marker in the ever-changing rhythmic dance of the universe. ***Events like the Big Bang are the bass notes and kick-drum beats of the infinitely repeating and rhythmic song of beingness; they help pump the deep cosmic groove.***

ANTIMATTER EXPERIMENTS

Antimatter represents another enormous gap in our understanding of the cosmos. *Antimatter* is fully physical, in that it actually has been found in our universe. Physicists have "seen" *antimatter* produced in high-*energy* experiments; traces from small and short-lived bits of this material have been detected after large, *energetic collisions* in *particle accelerators*. *Antimatter* particles are built like those of normal matter, except they hold the opposite charge. *Antimatter* parallels *matter* in every way, and it is believed that all subatomic particles have their equivalents in *antimatter*. There could be entire *galaxies* identical to ours, but instead of being made of *matter,* they are made from *antimatter*.

When *matter* and *antimatter* come into contact with each other, all the *mass* of both is converted in an enormous and complete release of pure *energy* that leaves nothing behind. Since these interactions free all of the bound *energy,* such encounters would be magnitudes more powerful than even *atomic* and *thermonuclear reactions.*

So far, these experiments have only produced very small particles but, as I write, physicists are trying to generate larger quantities of *antimatter* in *supercolliders;* and they have no absolute certainty about what will happen if and when they do. It is noteworthy that, similarly, scientists also lacked certainty about what would happen when they exploded the first *atomic bomb.* In the early 1940s, some physicists even speculated that there was a theoretical possibility of starting a chain reaction that could spread and annihilate the entire Earth, yet they still went ahead with the project! The possibility of world annihilation does not seem to be an effective deterrent for curious or driven human beings. Maybe, deep down, we all understand that this physical world is just an *illusion,* so we are not so concerned with the small details, such as the annihilation of an *illusion.*

Based upon what we understand today, *antimatter* and *dark matter* are not directly related. *Antimatter* is physical, real, and detectable, and we understand how it reacts with *matter. Dark matter* is theoretical, yet it seems to be much more plentiful than even *matter.*

EMPTY SPACE AND VIRTUAL PARTICLES

As mentioned, "empty" space is not really empty at all. It was discovered recently that space is filled with an endless succession of very short-lived particles, fluctuating in and out of existence. Because these *particles* only exist for the briefest of moments, we usually refer to them as *virtual particles.*

Looking closely at *protons*, we now know that "empty" space makes up 90 percent of their *mass.* This "empty" space has actual *mass,* and some physicists theorize that the *mass* comes from these *virtual particles.* Since we are made from atoms, our bodies also are made from these same *virtual particles* that continuously move in and out of existence.

There were good arguments for relating *virtual particles* to *dark energy* since "empty" space contains both. In the attempt to establish this relationship, the *mass* of *virtual particles* was calculated, but this calculated *mass* was found to be far too large to explain *dark energy.* This calculation was not just a little off; it was off by the unbelievably huge factor of 10^{120}. This enormous discrepancy was so dramatic that some called this the worst prediction in the entire history of physics. It seems clear that *virtual particles* are not directly responsible for *dark energy.*

Even though the equations of physics describe both *particles* and *virtual particles* equally, physicists had not been able to see, touch or measure any of these *virtual particles,* despite the existence of many observable physical phenomena that result from, and even require the involvement of *virtual particles.* Today, their "virtual" nature is being re-examined because one form of these *particles, virtual photons*, has just been "captured" like "real" *photons* and then used to create *light*. The shifting of other *subatomic particles* from *virtual* to *real* is now seen as imminently possible due to this recent discovery involving *virtual photons*.

Virtual particles are just one more "thing" out there that we really do not understand, but they are of special interest because they seem to be "located" somewhere near the cusp, or outer-edge, of our current understanding. *Virtual particles* could be very well positioned to become the next new window into the wonderland that lies beyond our three-dimensional *viewport*.

WHEELER, PENROSE AND BLACK HOLES

John Wheeler, who ended his career at the University of Texas, Austin after many productive years at Harvard, and Sir Roger Penrose, of both Cambridge and Oxford, developed much of the physics that describe *black holes, wormholes,* and other unusual topological features within *spacetime*. They, and now others, have assembled an enormous body of work describing these features; they all derive from *relativity* and can be understood mathematically. *Wormholes* are still only theoretical but physicists have substantial physical evidence of actual *black holes,* and they are constantly discovering new ones scattered about our universe. As I write, cosmologists have just discovered two new, enormous *black holes*, one of which is 20 billion times the *mass* of our Sun.

Black holes are described as enormous deformations of *spacetime* due to an ultra-dense *gravitational mass*, usually a collapsed star. They are often found near the center of *galaxies* and, while massive, black holes are also very small in the terms of three-dimensional space. *Gravity* becomes so great in these *cosmological events* that nothing, not even light, can escape its pull.

On the other hand, *wormholes*, which are potential bridges between two remote areas of *spacetime,* are still theoretical. It has even been seriously proposed that *wormholes* might function as physical doorways to *parallel universes*.

Due to their enormous size and nature, these powerful structures involve extremely large and, therefore, extremely interesting *energies*.

They represent the cosmic storms: the typhoons, hurricanes, and tsunamis of our local universe. These events also provide a unique and revealing window for a new level of understanding.

A recent and very interesting theory about *black holes* proposes that *holographic information* describing our entire universe could be mapped onto their sides; *black holes* might be the actual nerve centers and brains of our physical universe. If this theory turns out to be true, it is conceivable that *black holes* might store and process all the *information* needed to create and sustain our three-dimensional universe. It might even be possible that they are a component part of a system that *projects* this *information* through the dimensions so that our universe only appears to exist, much like a *hologram*. Creation only gets stranger the more we try to understand it from our three-dimensional perspective.

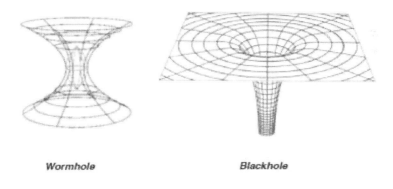

Wormhole   Blackhole

NESTED STRUCTURE OF THE "HOLOGRAPHIC UNIVERSE"

Throughout the known three-dimensional universe, extending from deep within the tiny quarks to far beyond the most enormous galaxies, a repetitive and consistent physical pattern seems to rule. Bigger "things" are constructed from mostly empty space with sparsely spaced, smaller, physical "things" that are all held together by various forms of energy. When we examine these smaller "things" closely, they reveal that they are made up of even smaller "things" that are distantly spaced and held together by energy. Each smaller piece is roughly modeled after the larger piece that it is part of, and vice versa.

Quarks, atoms, molecules, solar systems, galaxies, and the universe are just some of our current names for these different levels of structured physical systems. We do not yet know if there are large or small limits to this pattern. It is entirely possible that as our technology evolves, our paradigms shift, and we begin to dream new possibilities, we will discover that this pattern does continue forever in both directions.

Such repeating patterns evoke physicist David Bohm's groundbreaking *holographic paradigm*. His theory proposes that the universe is constructed in a way wherein every smaller part contains all the necessary and same information as all the larger parts. Information is nested so that each and every piece contains all the information necessary to recreate the whole.

More recently, and specifically, this idea has been expanded into the *holographic principle*, which states that all the information contained within a volume of dimensional space can be contained on the surface boundary of the region. As mentioned, there is speculation that *black holes* might be extremely dense storage systems for all the information in our universe, meaning that the history of everything in creation might be fully mapped on the surface of *black holes*.

We are familiar with the visual and visceral power of a well-constructed *holographic projection*, having witnessed dramatic examples at museums and technological art shows. The current ultimate in three-dimensional imagery, *holographic images,* can be so dimensionally life-like that we can physically walk around them and even observe them from different angles. Until we actually try to touch them, well-constructed *holographic images* seem quite real.

Holographic images are generated by splitting coherent light beams, which are formed from a single frequency of tightly focused, *Laser light*. One-half of the split beam, the *object beam*, is projected at the object that is being reproduced. This beam is then scattered, by reflection off of the object, towards a special photographic plate that is similar to a film negative. The other half of the split beam, the *reference beam*, is projected directly onto this photographic plate. Once the two halves of the single beam rejoin, the differences in these two parts form an *interference pattern*, which is then permanently recorded onto the photographic plate. To our eyes, this recorded pattern looks random or even abstract, but when the same *frequency* of *Laser light* is projected through the plate, a life-like duplicate image of the original three-dimensional object magically appears.

An extremely interesting property of this type of *holographic image projection* is that if we take the flat negative, break it into smaller pieces, and then project the *Laser* through a single broken fragment; we do not see only a part of the *image* as we would expect to see with a standard photographic negative. Instead, we see the entire *image*, although it may be less sharp-edged, less bright, or less dimensional. While it may be less detailed from particular angles, each broken piece still holds enough information about the whole to create a lower *resolution,* yet complete, *image*.

As scientists learn more about this type of *holographic storage*, they are simultaneously uncovering another interesting aspect that involves the cross-dimensional transfer of *information*. For example, we know that when the actual *holographic plate* is only two-dimensional, it is still able to project an image that has some of the qualities of a three-dimensional image. Such a well-constructed holographic image appears to occupy, at least partially, three-dimensional visual space, even though its information is only stored on a two-dimensional surface. Here the *hologram* is functioning as a type of bridge between two and three-dimensional space.

Considering what we already know about *holography*, I find myself wondering about other possibilities. If, instead, a three-dimensional "negative" becomes the source, will it, then, project a four-dimensional *image*? What if we expand this concept to five dimensions, or even eleven dimensions? Could it be that our solid-feeling, three-dimensional world, exactly as we experience it right now, only feels physical because it has been created and "solidified" through multiple levels of *holographic imaging*? As discussed, some cosmologists have proposed that *information* describing our physical universe might be stored *holographically* on a particular type of "flat" surface that acts like a *holographic plate*: the side wall of a *black hole*. As dimensional levels compound and the amount of raw *information* increases, do these *holographic images* begin to take on a more "solid" feel? If the plate originates in eleven-dimensional space, does the eventual *projection* into three-dimensional space actually become physical? If this turns out to be true, then we may be living in the ultimate example of a solid-feeling, three-dimensional *image*: our universe—*projected* through *information* generated and stored elsewhere in the *Multiverse*. The solidity of our world might only be the result of multiple layers of *projected information* from deeper dimensional levels; our world, while appearing solid, might really only be a *holographic image*. Could this be the actual mechanism of the illusion or *Maya*, from ancient Hindu texts?

What might it mean for our universe to be both *holographic* and multi-dimensional? Recall that flat, two-dimensional holographic plates store all the information necessary to create three-dimensional *holographic* images. We might then extrapolate that each lower dimension holds and stores all the information for the next "higher" dimensional *projection*. This would mean that all the information stored in any "lower" dimension would be, in turn, contained, filtered, and then stored, again, in the next "lower" dimension.

With this type of cross-dimensional storage, even one-dimensional space will contain all of the necessary *information* to describe the entirety of creation. There is no *information* missing from any dimensional realm, including our three-dimensional *viewport*. Instead, *information* is just dimensionally "flattened" or filtered, making it impossible for us to directly visualize and understand. While we experience only a small slice of the full Multiverse, it still contains all the *information* that is necessary to describe the entire Multiverse. ***This means that everything within creation is already right here and available; contained within our own bodies is everything that is, ever was, and ever will be! We are deep within a journey of self-discovery—one that is about opening, evolving, and learning how to access and use this infinite amount of information.***

IT IS ALL ONLY INFORMATION

When physicist Brian Greene asked John Wheeler what the most important topic of physics will be as we deepen our explorations, Wheeler's response was ***"information!"*** At the most basic level, he saw the very real possibility that the most elementary thing in the universe is simply *information*.

This *information* then expresses itself as the very *particles* that organize themselves, through patterns of *vibration* spanning multiple dimensions, to form the entire physical structure of our universe. ***Patterns and overtones of vibrational information freely communicate and interact through all dimensions, thus creating*** *the entire order and structure of our universe. In our three-dimensional reality we do not see or experience this raw vibrational information; we only experience the "shadows" that eventually reach our three-dimensional realm, and only after information has been "distilled" from eleven or more dimensions into our familiar three dimensions.*

A PERSONAL PERSPECTIVE ON QUANTUM WEIRDNESS

Q*uantum physics* has given us an entirely new way of describing physical interactions at the level of small subatomic *particles*. We have discussed the *Many Worlds* interpretation in which all possible outcomes can be viewed as actual realities that occur somewhere beyond our senses, possibly in other *parallel universes.* We discovered that there is always some *probability* that anything "allowed" could happen. With the right *energetic* conditions, when two *masses* meet all *particles* could pass through the vast spaces that exist between particles, rather than collide. In the *Many Worlds* view, this normally improbable outcome would occur in an infinite number of *parallel universes*. If we could somehow instantly shift our awareness to one of those other universes, we would then be able to walk through walls, or even "walk on water!" We need to consider the possibility that it is our deeply ingrained patterns and conceptual ideas and conditioning about the solidity of our universe, including what we believe is possible, that keep us from participating in these different types of interactions.

As the reader knows by now, my personal belief about the meaning of all this, while not unique, is still different from most mainstream explanations. I have the clear sense that it is largely our brain's need to time-order the unfolding of reality, which inhibits our ability to see the greater truth. "Time" implies speed and distance; so, by its very nature, it creates separation between things and events. For example, I believe that when we observe *non-local* behavior between *paired particles*, it is not because some infinitely fast *information* pulse or wave is connecting the two particles, as David Bohm theorized. Rather, it is more likely that this instantaneous transfer occurs because they were not really two different *particles* to begin with; both *particles* are unified at some deeper level of existence. No matter how far apart the pieces appear from our perspective, a singular oneness exists at the core of creation. ***The real meaning of non-local behavior is that everything in the universe is always completely***

111

and deeply connected. The appearance of separation and "time" passing is only the result of the way our three-dimensional brains need to separate, organize and view information.

The "odd" *quantum experimental results* are simply perceptual anomalies, created by viewing and conceiving creation in fewer dimensions than its original source. The difference between the old black-and-white movies and the three-dimensional version of *Avatar* only hints at the kind of spatial transformations that we will discover as we unlock this secret.

In our conversations, we habitually attempt to explain multi-dimensional phenomena using terms and ideas solidly grounded in our current paradigm. For example, when we try to imagine what it would be like to occupy other dimensions, we still talk or think about "time," distance, and shape. Our conceptual brains and our language cannot walk us beyond these old ideas. To really understand multi-dimensional space, we need direct experience, and this requires our evolution and expansion.

Evolution is a process; it does not happen all at once. A practical first step is relaxing our ties to and need for our old concepts; this practice makes room for new ways to experience creation. To do this, we must first recognize which ideas are anchoring us firmly in our old *viewport.* *Freedom involves knowing how and when to let go of the very helpful concepts that have been serving us so well.*

"TIME"—THE COSMIC TRICKSTER

We often include our human invention of "time" when describing how things appear to us. The sphere that visited Flatland was observed as a circle that changed size with "time." "Time" is the wild card, the joker, often used to explain phenomena that otherwise would make no sense to us. "Time" seems very important in our three-dimensional worldview, but really, it is just a terribly misunderstood and orphaned trickster.

We sometimes perceive "time" as our enemy: the constantly marching, relentlessly advancing quantity that defines our limited stay here on Earth. We imagine that if we could only stop the advance of "time," we would not age or die; we could become immortal.

Digging deeper, we discover that there is nothing really to change about "time" except our attitude. "Time" actually does not "march on" or exist as an absolute. Instead, we create the concept of "time" to facilitate our three-dimensional expression of life.

"Time," as we know it, is a necessary conceptual tool for living in our three-dimensional reference frame. "Time" and mind are completely interrelated. In the words of Einstein, "Time is a stubbornly persistent illusion." In the words of Eckhart Tolle, "Time and mind are inseparable. To identify with the mind is to be trapped in time."

Spiritual leaders and scientists completely agree that our perception of "time" is an illusion. In physics and math, we can think of "time" as being similar to other dimensions. In mathematics, we might find ourselves referencing the future and the past in much the same way we might look to the East or West. In physics we might say **the present moment informs the "past" and the "future" equally through waves of information that travel in all directions.** Our more human perception of "time" is usually as if we are stuck on a freeway heading north, not realizing that we could easily slow the car, or even turn around and head to the south. We only understand "time" from the perspective of a car that is always traveling at 70 miles per hour in the northbound lane, a very limited perspective. We just cannot fathom the possibility of slowing down or stopping the march of "time," or making that U-turn to travel the other direction.

Possibly, if we were traveling in a multi-dimensional hovercraft, we could also travel in a different direction and leave the realm of "time" completely, in much the same way flying liberates us from our restricted, two-dimensional, ground-level contact with the Earth's surface. To understand "time" differently we must shift paradigms. *The very first step of this process is to better understand how and why our current sense of marching time limits our vision. We must shed our old, restrictive ideas about time if we are to allow for the emergence and growth of a new vision.*

MODELING THE PAST, PRESENT AND FUTURE

A simple demonstration can help us visualize "time" differently. Imagine that we are on a rapidly moving train. Looking out the side window, we see the landscape as it passes. Having just finished our lunch and used the last paper towel, we notice its cardboard tube begging to be used in some creative way. Playfully, we put the tube up to our eye and look out of the side window with only that one eye open. Pinning the tube against the window, we now see the small portion of the view that happens to be framed by the tube.

Think of the landscape we see through the tube at any given moment as the "present." We cannot yet see the "future" landscape until it pops into our tubular view. We remember, but can no longer see, those scenes that have already moved through our field of vision; those remembered scenes are our "past." This tubular view becomes our "world," and we discover that we can make reasonable guesses about the "future" scenes based on our "present" landscape. As the train moves forward, bringing new parts of the landscape into our view, each next "view" becomes our next "present." "Time" marches forward relentlessly, and if we forget that we are on a train, this linear and one-directional "passage of time" becomes our world and our reality until, playfully, one of our kids knocks the cardboard tube away from our eye. ***Suddenly, right in front of our eyes, is the "past," "present" and "future"—all at once and fully interconnected!*** The once divided and ordered "past," "present," and "future" now form one continuous landscape that could be traversed in any direction.

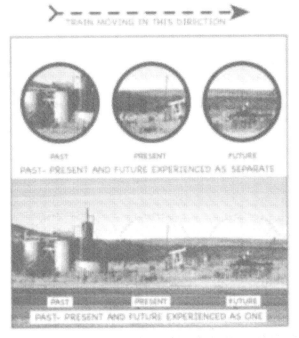

Looking out of a moving train window through a cardboard tube, we can only see a small part of the view at any moment. We remember the "past," see the "present" and expect to meet the "future." When we remove the cardboard tube which acts as our visual blinder, the "past," "present" and "future" can then be seen as one continuum.

"Time" is the way our brains divide, organize, catalogue, and store an otherwise *infinite* continuum of *information*. **Time is our loyal assistant, preventing everything from happening at once and overpowering our limited processors.** Time is nature's way of intentionally restricting our *"viewport"* into the wider universe: a necessary and beautiful constraint that helps shape our three-dimensional specialization. "Time" is the ordering system that allows us to form and organize our conceptual thoughts. *We are not built to handle everything unfolding at once. For information to be useful, it must be processed by our brains in smaller increments. We stack and order these increments by creating something called "time." "Time" is the way our brain organizes information—it is our brain's filing system.*

RESONANCE

Resonance is an extremely important phenomenon. The term comes from *wave physics,* and because we observe it every day in many ways, it is something that we can easily understand but often take for granted. *Resonance* is defined as *"the intensification and enriching of vibratory phenomena by supplementary vibration."*

Vibration produces *waves.* When two *waves* of the same type meet in the same medium (air, water etc.), they interact (*interfere*), either destructively *(destructive interference)* or constructively *(constructive interference),* or more typically, in some combination of these two. In music, certain *frequencies* or notes combine to create a more *resonant* whole—a fuller harmony. When two waves of the same or related frequency (pitch) vibrate *in-phase,* the waves will add together (*constructive interference*), and we say they *reinforce* each other. This new, combined wave will have an increased *amplitude,* and if these *waves* are *sound waves,* they will sound fuller and louder. If these same *waves* (*sound waves*) are *out-of-phase*, meaning that the directions of the *waves* oppose each other, then they will cancel each other, resulting in the full or partial destruction of the *vibratory wave.* This is *destructive interference,* the common vibratory phenomenon that noise-canceling headphones utilize.

The mixing of *constructive* and *destructive interference* helps form and shape the music, language, noise, and everyday sounds that we all hear. *Feedback* in a PA system is a dramatic example of extreme, uncontrolled *constructive interference (reinforcement).* A beautifully sung harmony is partially the result of carefully controlled *reinforcement,* while the discordant *diminished seventh* chord is an intentional mix of *reinforcement* and *destructive interference.*

All physical things in our world have one or more *natural resonant frequencies:* the frequency at which something will vibrate naturally and easily. When we strike a drum or a piece of china, the note that we hear is its *natural resonant frequency.* When an object interacts with a wave that is also oscillating at the object's *natural resonate frequency,* the object may start to *vibrate* at that same *frequency* or at some multiple of that *frequency.* **Resonance is the term used for describing the reinforcing relationship between two or more things that are vibrating together.**

Understanding Resonance

Humans have a deep, natural, and physical understanding of vibration. Sound waves, electromagnetic waves, and ocean waves (including tsunami, earthquakes, and more) are all formed from different types of materials in *vibration*. Since sound is a very familiar and relatively simple type of *vibrational wave,* which we all experience, we will use it as the example.

A guitar string is a convenient way to begin our discussion of basic *sonic vibration* and *resonance*. Pluck the low E string and watch it vibrate. The *frequency* is the number of times the string makes one complete back and forth trip in a single *second*. For example, the *frequency* of the A string on a bass guitar might be expressed as "110 vibrations per second"— also referred to as 110 Hz (Hertz), or sometimes 110 cycles per second (CPS).

The musical term for *frequency* is *pitch* or *note*. If we hit two strings side by side and they both vibrate at exactly the same *pitch*, say 440 Hz (modern "A" pitch), we say they are in perfect *resonance*. The combined tone is *reinforced,* therefore it sounds much louder and fuller. If one tone is 430 Hz and the other is 440 Hz, then when they vibrate together they will compete with each other and produce a third vibratory sound, called *interference* or *beats,* which represents the difference between these *frequencies*. In this particular example the difference in frequencies is ten, which results in a ten-beat-per-second, audible vibration—a Wah-Wah sound that most guitarists seek to eliminate when they tune their instruments. *Harmonic resonance* also happens with halving or doubling a *frequency*, and we call these special cases *octaves*. Lesser *resonances,* called *harmonics,* also occur at multiples and divisions of 3, 4, 5, 6, etc. A trained guitar player knows the exact places on a guitar string to "tap" and create these special sounding, *resonant* tones.

Sympathetic resonance occurs when a secondary object begins to *vibrate* on its own because it has the same *natural harmonic frequency* as the original source *vibration*. For example, another string or the body of an instrument could start to *vibrate sympathetically* if it has the same or related natural *resonant frequency* as the string that originally was plucked. The physical *vibrations* from the first string are transmitted directly through the air and to the *vibrating* body of the second instrument. *Secondary sympathetic vibrations,* which arise within this and other instruments, are then transferred to the ears and bodies of the audience.

This effect can be intentional, as it is with well-tuned symphony orchestras, or it could be accidental and undesired. In 1940, the wind blew through and *vibrated* the Tacoma Narrows Bridge just like the reed of a woodwind, creating a howling sound and so much *sympathetic resonance* that it caused the bridge to fly into pieces. This collapse, which you can view on the Internet, taught engineers about the need for *damping* in bridge design; as a result, modern bridges are designed to prevent this physically destructive type of *sympathetic resonance*. If a musician accidentally leaves an electric instrument near their amplifier and walks away, an electronic version of *sympathetic resonance* can build into a deafening sound called *feedback*. Sometimes, in rock and roll, punk, and contemporary electronic music, this *feedback* (amplified *sympathetic resonance*) is created intentionally as part of the music. Jimi Hendrix was famous for utilizing this principle. Controlled *sympathetic resonance* can also be built into the instrument itself, as it is with the *secondary vibrations* of drone strings on a sitar, dulcimer, or similar instrument.

Musical instruments are always communicating with each other as their component parts exchange *energy* and *information* in the form of *vibration*. Everything that we call music is simply the controlled interaction of sonic *vibrations*.

We respond to and *resonate* with many types of *vibrations*. Some, such as *visible light,* we can see, while others we both see and feel, such as the waves in the sea. Sound waves are heard but cannot be seen, while radio waves and microwaves are unseen, unfelt, and unheard by our bodies, but not our scientific instruments.

String theorists believe that matter begins with *vibration;* because we also are made from matter, our very existence is a sign that we are responding and *resonating* with some parts of this deep, original, source *vibration*. **Our natural resonant frequency at any given moment determines just what part of this source vibration we respond to most fully. As we evolve, we change and expand our natural resonant frequencies, which also changes the way our world appears. As our individual patterns of vibration change, the world we each experience will also change.**

RESONANCE THROUGH MULTIPLE DIMENSIONS

It now seems that everything that we know and consider to be real in our world is formed from vibration at some fundamental level of creation. This original symphony is what makes all particles and matter. The particular forms that "manifest" in our universe and our individual lives are the result of our ability to resonate with these particular parts of creation. Our personal universe is created and experienced through selective resonant vibration.

We can think of the universe as a giant, vibrating, musical instrument (*M-theory*). **Sympathetic resonance producing harmonics that connect through multiple dimensions is the way the universe creates and organizes itself.** Vibration at the very core of creation creates a "symphony" that is infinitely rich with *harmonics* and other *vibratory information.* This complex *vibratory information* then casts its "shadow," creating appropriate forms of *sympathetic resonance* throughout all dimensional layers. **The resonant "shadow" of this deep-level "symphony" is eventually perceived locally as the physical manifestation of our universe.**

Everything in creation is intimately and directly connected (*non-local, enfolded*), so we are always communicating with the deeper levels of creation through our *vibrational* state. Our *frequencies* and *harmonics* form our personal "song," which then, in turn, *resonates* with all other parts of creation. **This is how we connect and communicate through the dimensional barriers with everything in creation. This multi-dimensional vibrational interconnection is what manifests and connects everything in every parallel universe. How we each vibrate determines exactly which parts of all these parallel universes we resonate with, and this, in turn, immediately determines how the world appears to us. We "understand" other dimensions and other dimensions learn about us through vibration—this continuous and fluid communication travels both ways.**

"Going with the flow" is simply allowing the natural resonate frequencies to vibrate us and move us along by fully utilizing sympathetic resonance. As our vibration shifts, our location in the universe will also shift, as we travel fluidly and easily (dance) through a universe that is always supporting us with this natural, sympathetic resonance. We are in essence just "hitching a ride," letting the universe do the "work" for us. The "path of least

resistance" involves learning how to allow our natural frequencies to resonate fully, so that we can enjoy the energetic "help" of the universe.

WHAT IS FREQUENCY OUTSIDE OF TIME?

All of our *frequency* discussions have included, by definition, the human concept of "time." *Frequency* is a three-dimensional, "time"-based concept. Since it is measured in *cycles* or *vibrations per second*, "time" is a necessary part of the description of any *vibratory* phenomenon occurring in our physical realm. *String theory* predicts that the fundamental nature of the universe is *vibratory,* but its ten- or eleven-dimensional space is almost surely not subject to our concept of linear "time." Within this type of space, *frequency,* and *vibration* must have very different meanings than they do in our three-dimensional world. At the very source of all creation, a place where there is no beginning and no end, there must exist a timeless form of *vibration*.

What would *frequency* be without the measurement of time? Such questions highlight the critical and important limits of our three-dimensional existence. We simply do not have the conceptual tools to comprehend *vibration* outside of "time." However, we do have the ability to feel or sense *resonance* from these deeper levels without relying upon our normal conceptual thinking. **The development of this type of sensitivity to "presence" is an important step in unlocking our ability to experience and participate more fully in the deeper dimensionality of creation.**

So what is *vibration* when it originates "outside of time," and what changes as our awareness of *vibration* becomes liberated from the constraints of concepts and time? Like *Zen koans*, pondering these questions can help us open up to new ways of experiencing and understanding our deepest interconnections.

In the next two sections, "Spirituality" and "Experience," we look at how our traditions, culture, and personal lives support this expanded vision of the universe

SECTION TWO–THE SPIRITUAL SEARCH

RELIGION OR SPIRITUALITY

Spirituality and religion may seem very different, but at the deepest and most profound level, most of the world's major religions are actually very similar to each other and deeply spiritual. Spirituality – the inner, personal search to discover the essence and purpose of one's own *being* – can be found deep within the heart of every major religion.

Today, organized religion and spirituality are often viewed quite differently because the public face of many of the world's religions can appear prescriptive, indirect and dogmatic. Major religious organizations typically focus significant resources on the organization's corporate, business, and political interests; these powerful and very influential sub-organizations have their own agendas, which may not always mesh perfectly with those who focus deeply on their religion's core message.

While most adherents may not be exposed to this core esoteric level of their spiritual practice, they still participate in their chosen religion because they desire a deepening of their personal spiritual experience—almost everyone is searching for meaning in their lives. The desire to understand our purpose and relationship to this world is fundamental to our evolution. For many of us, organized religion provides a safe, structured, socially acceptable framework for this deeper exploration. Therefore, when I speak of spirituality, I am including all of the various ways that human beings have used to search for meaning and connection in their lives.

SCIENCE AND SPIRITUALITY: TWO SIDES–SAME COIN

"All religions, arts and sciences are branches of the same tree. All these aspirations are directed toward ennobling man's life, lifting it from the sphere of mere physical existence and leading the individual towards freedom."

Albert Einstein, *Out of My Later Years-1956.*

Science and spirituality both seek to understand and explain the most fundamental, yet often hidden, truths about life. Until recently, these two approaches often appeared to be completely separate, competitive, and often even contradictory pathways to truth. It makes logical sense that, as both of these primary human endeavors evolve, they will tend to expand and emphasize their similarities, since both are directing us to look towards the same objective: a deeper truth. The confusion or conflict between religion and science that we experience today is only an illusion created by our individual separation and limited perspective. Whatever we are, we are not multiple things with multiple sets of laws governing our existence. Eventually, spirituality and science must merge because, ultimately, everything is deeply interconnected. As our understanding evolves, these two great human endeavors will be understood one day as nothing more than two different views of exactly the same thing. Today, as this old dichotomy is rapidly disappearing, many physicists are beginning to sound more like ancient religious mystics. At the same time, there are many religions and spiritual practices, new and ancient, which speak of a more personal and direct experience with the infinite possibilities of *Being;* some even directly incorporate aspects of these new and "crazy" ideas from physics.

Still, even with all our technical wizardry, this same science has not been able to explain the most fundamental truths about our human experience. Despite all the theories and analysis, *life-force*, the primary and critical seed of life itself, still remains unexplained and mysterious. Scientific voices in every generation have boldly declared they were on the cusp of finding God in the machine, but no efforts in this direction have ever produced a clear understanding of the fundamental source of life itself: our spirit. Good answers to these fundamental questions have been completely lacking and, until now, many of the more resonant answers have come from the spiritual side of our explorations. Because these questions and answers are always framed by our old, three-dimensional concepts, we will always be limited and critically flawed in our ability to describe the real workings of the multi-dimensional *Multiverse*.

DIFFERENCE BETWEEN FAITH AND BELIEF

At the very core of all real personal and social change are our deeply held individual beliefs. As Gandhi said, "Your beliefs become your thoughts, your thoughts become your words, your words become your actions, your actions become your habits, your habits become your values, and your values become your destiny." Henry Ford famously said, "If you believe you can or if you believe you can't, you are right." Henry Ford and Gandhi both understood something quite profound – that our personal beliefs are where we define our lives; they contain and shape who and what we become.

For the purposes of this book, I will define a critical difference between *belief* and *faith*. *Belief* requires an element of proof or verification acquired through personal experience. The validation might be scientific, or vague and experiential, but *belief* is always built upon some form of personal validation. *Faith* does not require proof; instead, it asks for acceptance without this type of personal validation. When *faith* includes the personal experience of validation, it becomes a *belief*. Both *faith* and *belief* define and shape our individual and collective lives.

I have built most of the discussions and conclusions of this book upon my deeply held *beliefs;* they are founded within my experience, but are also supported by our best and most recent science. These ideas do not begin as *beliefs*; in my personal process, I sometimes experience a brief moment of inspiration, which is a signal to wait patiently for additional forms of verification. This waiting requires a focused yet open form of heightened attention that can take moments, months, or sometimes even years. In my books, I have attempted to identify clearly the ideas that are still "leaps of *faith*"—directions of thought that I have yet to personally validate. As discussed, many of the ideas in this book are not scientifically provable because of our limited reference frame (*viewport*), but, for me, these are still *beliefs* because they are grounded by some combination of science, a cultural "knowing," and my personal experience.

The line between *faith* and *belief* can be confusing, since the two are often intermixed. This confusion happens naturally because a person's *faith* always informs their *beliefs*, and vice-versa. To create deep and long-lasting personal change, ideas must reside beyond our minds to *resonate* deep within the very core of our *being*. For this type of *resonance*, both the heart and mind both must be open and aligned with the rest of our *being*; divided, our heart and mind are each relatively powerless. *Faith* or *belief* originating in the mind can eventually expand to include the heart as we evolve, but when *faith* is still centered in mind alone, it lacks this deeper *resonance*. It is that which we hold in our hearts and support with our entire *being* that ultimately determines how and where we engage within our *Multiverse*.

ZOROASTRIANISM—ORIGINS OF DUALISM AND MONOTHEISM

A fundamental principle of many Eastern spiritual traditions is a recognition that the world is *dualistic* in its nature. *Dualism* is the belief that the world exists through the opposition of two forces or powers. The oldest recorded documentation of *dualism* as a formal philosophy places its origination in Persia within Zoroastrianism, also called Mazdaism. Zoroastrianism's roots may be much older, but its first recorded history involves the writings of the ancient Iranian prophet Zoroaster, or Zarathustra, approximately 3,000 years ago. Combining aspects of both dualism and monotheism, it is the single parent of many of today's major world religions. It is also the world's oldest organized monotheistic religion.

The foundational principles of the Jewish, Christian, and Muslim faiths, including heaven and hell, Satan, God, the soul, virgin birth, final judgment, and resurrection, all directly derive from Zoroastrianism. According to the Zoroastrian story of creation, *Ahuru Mazda* existed in light and goodness above, while *Angra Mainyu* existed in darkness and ignorance below. They have existed independently of each other for all time; thus *Dualism* in Zoroastrianism involves the existence, but complete separation, of both good and evil.

In this original expression of dualism, evil is not God's equal opposite; rather, Angra Mainyu is the destructive energy that opposes God's creative energy. Aging, sickness, famine, natural disasters, blights, death, etc., are all attributed to *Angra Mainyu,* who lies outside of God's creation; the negative is seen as an outside element that opposes God, instead of as an integrated part of God's creation.

With later and more mature forms of *dualism,* we observe a more balanced dynamic between life and death, day and night, good and evil; one aspect cannot be understood without the other because physical life will always be a mixture and balance of these opposing forces. Our world depends on these opposing forces; both the light and the dark are necessary for us to experience all the different possible degrees of illumination. The coexistence and balance of good and evil allow us to understand and experience the full range of human behavior. We can say something is hot only because we can compare it to cold. *Evil, cold, and dark are not simply elements assaulting our world from the outside; they are all equal and necessary components that help build and maintain the physical structure of creation.*

Our very existence is entirely dependent on this dualistic nature of all the things that form our physical world. Dualism is the critical component that allows our brains to function. Our brains require contrast because the nature of our conceptual thought process requires that all things be measured relative to other things. Dualism forms the structural foundation upon which we build our three-dimensional existence. *Contrast is what makes our lives possible.*

What is sometimes called *non-dualistic* awareness generally involves being able to see, feel, or understand in a different way. It requires a type of understanding that lies beyond the separation of *dualism*. In *non-dualistic* awareness, there is no thought, no logic, no comparisons, no rational arguments, and no convincing; it is a process that only exists beyond conceptual thought and three-dimensional form. To even consider it as a type of thinking associates it too directly with our normal, three-dimensional mindset. *Non-dualistic awareness* can be described as a way of being and feeling that is free of thought. Even though there are many spiritual paths based on teaching or practicing *non-dual awareness*, this way of being and experiencing has been marginalized in our modern world, especially in the West.

Within extreme *non-dualistic* religious and spiritual practices, we sometimes see a tendency for adherents to think of *dualistic* thinking as a problematic or a lesser form of thinking. By teaching that only pure *non-dualistic* thinking can lead to *enlightenment*, *dualistic* thoughts can even be demonized and presented as something for practitioners to avoid. This view is steeped in irony because its concept involves strong judgments that actually create even more separation; it presents the "very *dualistic* idea" of building more good (*non-dualism*) and discouraging the bad (*dualism*). In this context it does not make sense to judge *dualism* as problematic because judgement, itself, is a product of *dualism*.

Duality is what forms and contains the structure of physical creation—our wonderful human playground. Any time we see *duality* as the problem or "the enemy," we are being critical of our very physical existence and the critically important three-dimensional aspects of our *being*. This is the spiritual equivalent of a scientist cursing *gravity*. There are days when we all wish *gravity* were not so relentless, but without it our physical lives simply would not work. Since our conceptual thinking is necessarily built upon *dualism*, labeling it as problematic is a criticism of our fundamental nature and our entire physical existence.

While awareness of *non-dualism* is very helpful for our journey towards freedom, any movement beyond *dualism* must first embrace all aspects of *dualism* as critical and revered parts of creation. Every stitch in the knitted fabric of *Being* depends upon every other stitch. **Enlightenment is expansive and never exclusive, which means that we must embrace separation and duality as necessary components of creation!**

Models and further discussions involving the interplay of *dual* and *non-dual* aspects can be found in *The Architecture of Freedom*. Below is the beautiful and well-known Taoist symbol, the Yin-Yang, which expresses graphically the dynamic, yet always balancing, *forces* of duality.

The Taoist Yin-Yang symbol expressing the dynamic balancing of opposites that make up duality.

EASTERN RELIGIONS AND PHILOSOPHY

MAYA AND LILA

Maya is the ancient Sanskrit word describing the illusionary nature of our existence. Within the Hindu faith, our *dualistic* physical world is seen as an illusion; we never encounter the real environment and everything that we perceive is only a part of this illusion. Our physical experience exists only because of the way our minds interpret this illusion. The game of life that we lead in this *Maya* is referred to as the *Lila,* which translates as "play." Our lives are only a game involving *duality*, within which we willingly participate. This means that none of what happens in this life can be interpreted as real.

In its purest form, the Hindu vision is completely consistent with the physics and many of the ideas within this book. Of course, as with any organized religion, over time the human mind has systematically transformed this deep and pure idea into many doctrines, rituals, and practices which were often created to support the different human agendas that have defined its history. Through this process, many important foundational principles and beliefs have been lost or hidden within these added layers of religious trappings. However, at the heart of every religion that endures, there must also exist a deeper truth: a core philosophy more aligned with the actual form of creation. Again, at the deepest levels, all *resonant* ideas must converge towards their common root.

Referring to the physical world as an illusion is not to say that it does not exist. It just does not exist in the solid, material, and fixed manner of our direct experience. The mind forms this material illusion from the raw *information* shared throughout the *Multiverse;* we directly perceive only the shadows of this deeper level of *information,* and these shadows then appear as the familiar forms of our world. ***We all share the clear and accurate perception that our physical world exists. The great illusion is that we interpret our perception as the only reality.***

ATTACHMENTS

Common to many Eastern spiritual traditions is a teaching that our personal *attachment* to aspects of this illusionary and *dualistic* world directly leads to all of mankind's suffering. We become *attached* to the illusion of our bodies, our ideas, our toys, our money, and our creations because we have learned to think that all these things are real, solid, and critically important to our lives.

In the West we certainly have become very *attached* to our possessions; sometimes we might even think that our lives actually depend on our "things." In my architectural practice, I have heard clients say things like "I couldn't possibly live without a five-bedroom house with travertine floors." Our possessions can become closely associated with our sense of well-being and used by our *ego* as an "asset" to help define who we are; they can even become the measurement of our own sense of self-worth.

While we might own the fastest car, biggest boat, and largest house in the most expensive neighborhood, and even feel "successful" for a period of time, we probably will also experience more stress as we worry about maintaining this new level of "self-worth." If we happen to have a financial reversal and lose some of these prized possessions, we might then start to imagine ourselves as "failures" as we become very unhappy. We might also worry that others will judge us as "unsuccessful," even if we have discovered that we are actually enjoying our newly pared-down lives.

The issue is never our possessions themselves—they are neutral things—rather, it is our individual and cultural perception that we need them, that they are important. The problem is our *identification* with these things. If analyzed and understood, we can quickly see that this "need" for a bigger house or faster car is certainly not a survival issue. It then becomes easy to see that these things are not critical to leading healthy, productive lives.

Attachment to friends, family, and loved ones is a much more difficult issue to examine honestly and reconcile. We are completely immersed in the mistaken belief that we are these bodies, just as our loved ones are their bodies. ***From our cultural viewport, we usually understand ourselves only through our individual bodies, which are clearly separate, isolated, and short-lived. This direct association naturally leads to constant worry and a desperate clinging to our present form. It is, therefore, nearly impossible for us to understand that this illusionary, three-dimensional concept—the deep-seated belief that "we are our bodies"—is the critical misunderstanding that prevents us from experiencing the full potential and possibilities that life is always offering.***

For the vast majority of Western people, our *attachment* to our body is nearly absolute. While this belief is likely to persist in all cultures, many Eastern traditions try to teach about a different mindset. They teach that anything that can die, decay, break, or get lost is *impermanent* and, as such, is "not real." Our bodies fall into this category. In contrast, the things that are considered "real" are those things that never die or be lost, such as our timeless soul and *Being* itself.

Our attachments to things, people, ideas, thoughts, and emotions can and do form actual physical areas of resistance within our bodies, blocking the smooth flow of energy and information. Attachments often lead quite directly to associated worry, anxiety, and depression. We worry that we might not earn enough income to make the big house or car payment. We worry about what the neighbors are saying about our Christmas decorations or our new landscaping. We have concerns about the popularity of our children and their future financial prospects. We worry about our bodies, our children's bodies, and our friends' bodies. "Are they growing up fast enough?" "Are we growing old too fast?" "What if we become ill?"

These habitual concerns create emotional and physical tensions that eventually result in energetic and physical blockages (which are actually the same). As these blockages manifest in our bodies, they first typically express themselves as knots, kinks, muscle spasms, and weakness that can interfere with our motion, comfort, physical creative expression, and emotional well-being. If these early symptoms are not addressed, the blockages will continue to build, eventually resulting in crippling physical and emotional pain and, eventually, sickness. Once fully expressed at this physical level, these individual surface manifestations of buried tension are assigned clinical descriptions like carpal tunnel syndrome, sciatica, arthritis, diabetes, heart arrhythmia, anxiety attacks, and ulcers.

Our deepest *being* does not want pills or other forms of temporary or symptomatic relief. It is really shouting for the tension to be addressed at its source. Pain and disease are nature's way of signaling that your body needs your attention.

While our unexamined *attachments* will eventually express themselves at the individual level as physical disease, injury, or deep emotional "suffering," on the broader, societal scale, these same types of *attachments* lead to struggle and conflict between different cultures that eventually manifest as war.

Naturally balancing and self-corrective in its nature, life always creates experiences for us that will signal, reinforce, and deepen our awareness of any unhealthy blockages. The natural world is very helpful in this way, whether or not we appreciate this quality in the short term. The process is never personal; it is simply the result of a fully interconnected and interactive system restoring balance. Through this natural balancing process, the unfolding of life will always highlight our areas of darkness as we are simultaneously presented with new opportunities to expand and grow our enlightenment. Healing is a natural and organic process. If we are fully open to whatever presents itself to us in these moments, our healing will always unfold in a most unexpected yet wonderful and miraculous way.

ATTACHMENTS DAMP OUR VIBRANCY

While our *attachments* to something, someone, some time, or some way are a natural part of human life, they are also the ultimate reason that we have difficulty experiencing the full "flow" of life-force. When we are attached to things or concepts, our core *resonant vibration* responds to the weight of this resistance by *vibrating* with less *amplitude* (power or volume) and fewer *harmonics* or *overtones* (expression and expansiveness).

In classical physics, *damping* is the term used to describe the effect of a weight or *force* that reduces the amplitude of a *vibrating* object. Imagine hooking a bungee cord between two columns on a porch and then snapping it like a big guitar string. While it is still vibrating, throw a towel over it and watch what happens. This is *damping*. *Damping* is usually a good thing in bridges and tall buildings because it keeps them from swaying too much.

With our energetic bodies there are certainly times when a controlled amount of vibrational *damping* may help provide the necessary calm in our lives. When *un-damped*, however, the *resonant vibration* of life can, and will, swell into something that infinitely much greater; this unrestricted flow of natural energy can be felt as repeating waves of unbridled ecstasy.

Our *attachments damp*, or restrict, our experience and prevent us from discovering the freedom and ecstasy that is possible and always available. This freedom does not require that we avoid or eliminate the actual things or ideas to which we are *attached*; the actual person, idea, thing, or object of our *attachment* is never the problem. Instead, we just need to better understand how *attachment* works and how it weighs us down. ***What is really blocking our full vibratory potential is only our concept about this person, thing, or idea.***

These people, objects, and ideas to which we become *attached* are often the most beautiful and sparkling gemstones of our lives. These include our children, our homes and prized possessions, our accomplishments, and our lifestyles. Freedom does not require that we avoid or distance ourselves from these. Neither does it require a simple or ascetic lifestyle, although there are many unexpected insights to be gained by living with such simplification. Instead, freedom requires that we reach the point at which we are no longer *identified* by these things. When we are not *identifying* with them, we can enjoy, care for, and love them, but not have our joy or love be fully dependent upon them. Freedom is built upon a psychological state in which our peace, happiness, joy, and love are not a direct function of or dependent upon the very things and people that we enjoy so much. ***We can learn to enjoy and be grateful for these loved ones and objects, but, at the same time, understand that these do not define who we are.*** This may seem subtle, paradoxical, or even illogical, but a critical step along the path to true freedom is understanding that our liberation from *attachment* actually results in an experience of greater love and joy.

In my architectural practice, when I present early conceptual ideas to a client there is a very real risk that the client will get *attached* to some aspect of one of the designs. If that occurs, the project will then cease to evolve freely. It becomes bound by an energetic but very real expectation that constrains the free flow of ideas so that the creative process becomes *damped.* Having learned that this can happen quite easily, I now never present my ideas until they are fully developed.

Witnessing, experiencing, and understanding our *attachments* are all critical steps for the realization of real and permanent freedom. ***From a Western perspective, we could define attachment as holding fast to the temporary forms of this life, including our memory of the past and hopes for the future, in such a way that we disrupt the creative present moment and our natural flow through life.***

131

BUDDHISM AND DESIRE

Deeper awareness and understanding of earthly *desire* is an important element of many spiritual paths, but the need to free oneself from all types of *desire* is found at the very core of Buddhism. This principle is clearly expressed within its four noble truths, Buddha's primary teaching. According to Buddhist tradition, all suffering stems from our personal *desires* and, therefore, letting go of *desire* itself is the most critical element to end personal suffering.

My own experience has been that *desire*, by itself, does not directly lead to suffering; *desire* is not the fundamental problem. Instead, *desire* can be used to naturally add an exciting human dynamic that can energize and help us focus. If wisely channeled, our natural human *desire* becomes a powerful and exciting tool that can enhance the quality of our lives.

Desire can be seen as a natural, beautiful, and energetic source for facilitating human expression. *Desire* can originate from the subconscious depths, from the body, or from the brain. At the physical, biological level, *desire* drives the reproduction and continuation of our species. From the artistic depths, it can result in the passionate physical expression of creative beauty that humankind has helped bring into the world. If expressed through the mind, *desire* helps produce the interesting intellectual and scientific works that shape and energize our culture. When expressed through the heart, *desire* fuels a wide range of spiritual explorations and expressions. *Desire* initiates much of the wonderful beauty and creativity that has been and will be manifested through human expression.

If, however, through our *desire* we become *attached* to a specific outcome, object, or individual, then we will inevitably suffer from this *attachment*. At first glance, it might appear that the *desire* is creating the suffering, when actually it is only our *attachments* to persons, things or results that lead to this suffering. To avoid *desire* simply because it can lead to *attachment* is equivalent to avoiding automobiles because they can lead to accidents. It is bad driving, not the car itself, that causes most automobile accidents. Similarly, it is not the *desire*, but our *attachment* to a specific outcome that leads to our suffering.

Our *desires* will influence and stimulate our *ego*, which, by design, always wants to "control" outcomes. *Ego* is also not the problem; it is a very useful, built-in mechanism for the protection of our physical form. Our *attachment* to some specific outcome is what signals *ego* to try to control circumstances instead of allowing a natural flow. *Suffering* will only occur whenever there is already some level of *attachment* to an outcome. No matter how wonderful and lofty the personal goal may be, if we are *attached* to a result we will inevitably *suffer*.

Throughout our lives, the deepest learning happens through direct experience. Nature balances everything, so the *suffering* itself will eventually guide the sufferer towards deeper insights into the true nature of *desire*, control, and *ego*. We will eventually understand that it is the *attachments*, and not *desire* or *ego*, which form the chains that keep us bound to our unhappy or unhealthy situations in life. Once we learn to dance with the *flow* by understanding the real nature of our *attachments*, *suffering* naturally drops away.

This is nature's way! We are here to gain experience in all its different forms, and the natural world uses *desire* as a way of making sure we do just that. *Desire* is one of nature's wonderful and powerful facilitators for three-dimensional living because it helps us focus and motivates a fuller expression of life.

OUR ATTACHMENT TO FREEDOM

As taught within some Eastern traditions, such as *Advaita Vedanta*, the focused and specific *desire* for freedom itself is the greatest teacher because it propels us forward through our spiritual path. For many seekers, it has been the specific *desire* to be free of *suffering* that has guided and directed personal journeys. **This same Eastern tradition also teaches that one of the last steps in the process of "enlightenment" is, by logical necessity, letting go of this "desire to let go of desire."**

At some stage, seekers realize that their *attachment* to freedom is, by itself, an impediment, so those on this journey are eventually invited to explore this critical and very interesting threshold. **Ultimately, we must examine the very desire to be free, which has long propelled the spiritual search for so many. Any individual attachment to this**

or similar goals must be released as a integral part of the process of becoming truly free.

TAO

Taoism, pronounced Daoism, is the ancient Chinese life philosophy based on living in harmony with the Tao: the source of all existence. Quotes from Lao Tzu, the master of ancient Chinese Taoism from the sixth century BC, sound suspiciously like he was also well trained in *quantum physics*.

"The Tao that can be told is not the eternal Tao."

"Nature does not hurry, yet everything is accomplished."

"Mastering others is strength. Mastering yourself is true power."

"When I let go of what I am, I become what I might be."

"To the mind that is still, the whole universe surrenders."

"Music in the soul can be heard by the universe."

"At the center of your *being* you have the answer; you know who you are and you know what you want."

"A good traveler has no fixed plans, and is not intent on arriving."

"All difficult things have their origin in that which is easy, and great things in that which is small."

"By letting it go, it all gets done. The world is won by those who let it go. But when you try and try, the world is beyond the winning."

"The wise man looks into space and he knows there are no limited dimensions."

"From wonder into wonder, existence opens."

"If you realize that all things change, there is nothing you will try to hold on to. If you are not afraid of dying, there is nothing you cannot achieve."

"Life is a series of natural and spontaneous changes. Don't resist them—that only creates sorrow. Let reality be reality. Let things flow naturally forward in whatever way they like."

"The key to growth is the introduction of higher dimensions of consciousness into our awareness."

"The words of truth are always paradoxical."

QI AND HOLY SPIRIT

Qi (or Chi) is the Taoist term for "circulating life energy," and it is based on their deep insights into the *vibratory* nature of the universe. This same life-force is recognized in many other traditions, although by different names. It has been called Holy Spirit, Prana, Mana, Ki, Ashe, Shakti, and many other names, all of which reference the same idea – that *energy* or *presence* that enables us to "feel alive." It is this life-force that defines the difference between the living person and the "empty" corpse that might remain only a few seconds later. We do not understand this process, scientifically, and yet we all can sense the exact moment that life-force no longer animates one of our loved ones. Life-force is always present while we are alive, but at certain times it can become much stronger or weaker.

The following discussion of Qi, from "About.com," evokes the language of *quantum physics*:

> In its broadest sense, Qi can be thought of as the vibratory nature of reality: how at the atomic level, all of manifest existence is energy—an intelligent, luminous 'emptiness' appearing as this form and then that, like waves rising from and then dissolving back into the ocean. Our perception of solidity—of forms as fixed and lasting "things"—is just that: a perception, based upon habitual ways of conceiving of ourselves and our world. As we deepen in our Taoist practice, these conceptions and perceptions of solidity are gradually replaced by the perception of the world as being more like a kaleidoscope with its elemental manifestations in constant flux and change.

Qi originates in extra-dimensional space but we continuously experience its shadow effects throughout our world. We still lack the ability to see, measure, or understand *Qi* at any level, especially that of its origin. However, in the East they have developed numerous techniques for understanding, tuning, working with, and manipulating *Qi* within our realm. Some of these techniques, such as Qi-Gong, Tai-Chi, and Acupuncture, have made their way West and have even become established practices. Our local Catholic Hospital in Austin, Texas now has a large acupuncture clinic where multiple aspects of Eastern and Western philosophy and technology are inter-mixed and working well together. Worldwide, Eastern and Western practices of healing and religion are all blending and borrowing as they discover their common useful connections.

REINCARNATION AND ENLIGHTENMENT

Fundamental to Buddhism and many other Eastern spiritual philosophies is the idea of reincarnation. In this belief system, each of us is born repeatedly to have lifetimes of experience until that final lifetime when we finally reach Nirvana or *Enlightenment*. How we live each lifetime determines the shape and form of our next lifetime.

While this idea can generally be seen as compatible with the structure of the *Multiverse,* there are two important differences between the Buddhist system of reincarnation and the mechanics of the *Multiverse.* The first is that in the *Multiverse* events, such as lifetimes, do not unfold in a linear sequence because time does not operate in only one direction. The "arrow of time" is only an artifact of our three-dimensional perspective. In the *Multiverse* there is no such thing as a next lifetime because all lifetimes are unfolding simultaneously and interactively, outside of time, in parallel universes.

Perhaps the other primary difference is more important to understand. The Buddhist belief system engages a form of judgment that promises rewards or "merit" for "correct" behavior. There are "noble" truths and ethical precepts, such as "right" understanding, "right" thought, and "right" action. It is taught that "noble" and "right" behaviors help one gain merit and *enlightenment*, the ultimate reward.

Judgment and reward are human concepts and are not active principles within the *Multiverse*. The *Multiverse* has no allowance for "good" behavior or "bad" behavior because cause-and-effect drives everything within creation. There is never judgment of any kind because all types of experience and behavior are equally important for creation. By its very nature, *Being* creates all types of experiences, both "good" and "bad," because this experiential pathway is the only one that leads to true wisdom, balance, compassion, and freedom. Like hot and cold, "good" and "bad" only help to shape *duality*; they are both required for our existence.

Enlightenment, the end goal of many Eastern spiritual practices, including Buddhism, can be defined as "obtaining sustainable freedom while still living in the *body*." It is not a single event but, instead, it is a process, or way of experiencing life, that is initiated only through the deep wisdom and compassion acquired by embracing, experiencing, and understanding everything—all the possible flavors.

SUFISM AND RUMI

Sufism is a mystical movement within the Muslim faith that began soon after the time of Muhammad. Formed as a reaction to the increasing worldliness of the faith, this particular branch, established in Turkey, became famous for its poets and whirling dervishes. Its most famous and most published advocate was the 13th century Persian prophet Rumi, who lived deep within the quantum mystery. The following express Rumi's wisdom.

> "Out beyond ideas of wrong doing and right doing there is a field. I'll meet you there."

> "The wound is the place where the Light enters you."

> "Stop acting so small. You are the universe in ecstatic motion."

> "What you seek is seeking you."

> "The minute I heard my first love story, I started looking for you, not knowing how blind that was. Lovers don't finally meet somewhere. They're in each other all along."

"Don't grieve. Anything you lose comes round in another form."

"If you are irritated by every rub, how will your mirror be polished?"

"You were born with wings, why prefer to crawl through life?"

"Forget safety. Live where you fear to live. Destroy your reputation. Be notorious."

"When you do things from your soul, you feel a river moving in you, a joy."

"Yesterday I was clever, so I wanted to change the world. Today I am wise, so I am changing myself."

"Everything in the universe is within you. Ask all from yourself."

"Be grateful for whoever comes, because each has been sent as a guide from beyond."

"Let the beauty we love be what we do. There are hundreds of ways to kneel and kiss the ground."

Why do you stay in prison when the door is so wide open?"

"Appear as you are. Be as you appear."

"Take someone who doesn't keep score, who's not looking to be richer, or afraid of losing, who has not the slightest interest even in his own personality: he's free."

"And you? When will you begin that long journey into yourself?"

"Respond to every call that excites your spirit."

"This being human is a guest house. Every morning is a new arrival. A joy, a depression, a meanness, some momentary awareness comes as an unexpected visitor... Welcome and entertain them all. Treat each guest honorably. The dark thought, the shame, the malice; meet them at the door laughing, and invite them in. Be grateful for whoever comes, because each has been sent as a guide from beyond."

RAMANA MAHERSHI AND ADVAITA VENTANTA

Ramana Maharshi was an Indian spiritual philosopher and teacher who was alive at the same time as Einstein. Self-taught, his individually developed sense of time, space, and interrelationship often made him sound as if he had been fully trained in the philosophical aspects of *quantum physics*. While he refined his own independent line of intuitive reasoning, *self-inquiry*, his overall philosophy stood squarely on the shoulders of the ancient *Advaita Vendanta* tradition. Fathering a modern branch of this very old, traditional lineage, his direct protégé, Papaji, had a special talent for attracting serious Western students to India. Papaji was directly responsible for training many of our most influential, contemporary Western spiritual teachers of this ancient tradition and its many westernized variants.

From his mountaintop in India, where he spent his entire life, Ramana shared these deep contemplative understandings that contain more than a hint of the quantum wonder world:

"Investigate the nature of the mind and it will disappear."

"The universe exists within the Self."

"Apart from thought, there is no independent entity called world."

"When your real, effortless, joyful nature is realized, it will not be inconsistent with the ordinary activities of life."

"The universe is only an object created by the mind and has its being in the mind. It cannot be measured as an external entity."

"The world is an idea and nothing else."

"There is neither past nor future: there is only the present. Yesterday was the present when you experienced it and tomorrow will also be the present when you experience it. Therefore, experience takes place only in the present and beyond and apart from experience nothing exists."

"Even the present is mere imagination for the sense of time is purely mental."

"The feeling of limitation is the work of the mind."

"The ultimate truth is so simple; it is nothing more than being in one's natural original state."

Covering one eye with his finger, "Look, this little finger prevents the world from being seen. In the same way this small mind covers the whole universe and prevents reality from being seen."

"Reality lies beyond the mind."

"It is all mind."

"Environment, time and objects all exist in oneself."

"Good or bad qualities pertain only to the mind."

"The numeral one gives rise to other numbers. The truth is neither one nor two."

I find it extremely interesting that his lifetime unfolded at exactly the same time, and yet worlds apart, from the breakthroughs in *relativity* and *quantum physics*. **Advaita and quantum physics are as different in background and methods as any two schools of thought can be; yet the deepest conclusions from both methods of inquiry are remarkably similar because both sincerely strive to understand the same universal truth.**

EASTERN MEDICINE

Because Eastern medicine once seemed "faith-based" from our Western perspective, I have included the topic in this section about religion and philosophy. Currently, we are witnessing the rapid integration of Eastern techniques such as acupuncture, Chinese herbs, Qi–gong, Shitatsu, yoga, and meditation into the healing practices of the West. There exists little or no scientific explanation for how or why these methods work; but it is clear they do, and sometimes they work better than anything else. Not too many years ago, most of these techniques were ridiculed as superstitious; but today, many of these same methods are successfully employed in major Western hospitals.

I experienced a very serious, seven-year medical ordeal that finally ended with a single visit to a Chinese acupuncturist in 1975. After many Western doctors had proposed radical life-altering surgery, one young doctor, who had witnessed many successful acupuncture treatments while working in Vietnam during the war, pulled me aside to encourage me to try acupuncture first. He made me promise not to tell anyone about his recommendation for fear of risking his license. That single treatment resulted in a complete change of my health and a radically different understanding of alternative medicine.

Studies of the *placebo* effect reinforce the importance of attitude and *belief* as an integral component of healing, but something else is also occurring with these once strange-seeming Eastern techniques. For more about the placebo effect, see the *Architecture of Freedom.* In some way, which the Western mind does not understand, they are working to improve the balance and flow of systems that involve deeper levels of our being. These techniques address more than just the physical.

WESTERN SPIRITUALITY

THE WESTERN FAITHS

Despite the long history of struggle and war over their alleged differences, the three major western religious pillars— Christianity, Islam, and Judaism—are all quite similar. All three derive from Zoroastrianism and claim to be Monotheistic, although the Christian belief in a Holy Trinity (separation of unity to facilitate our understanding) is seen by some scholars as a form of Polytheism. The other major differences are that Judaism does not recognize Christ as a Prophet and Islam's most recent Prophet, Muhammad, roamed the Earth about 600 years after Christ. It is, of course, ironic that while all three claim to be religions of peace, there has been a continuous history of wars fought over their differences. The aggressive nature of these religions is always the result of man's "after-the-fact" misinterpretation (and manipulation) of the words and intentions of all three religions' prophets.

While these three institutions continue to struggle about how to define their reasons for, and degrees of, separation, there has also been a parallel but radical revolution in spiritual thinking. New and expansive ideas are being thoughtfully explored worldwide, with the result being a massive cross-pollination of ideas from Eastern and Western spiritual traditions and modern science.

CHRISTIANITY

I begin with and focus on Christianity only because it represents my first and deepest personal exposure to formal religion. Most of the principles discussed here also apply to the Jewish, Muslim, and related faiths because all embrace a similar set of beliefs and form. Seeking connection, not separation, I am personally less concerned with the differences for these have been argued for millennia.

The common Christian description of God has not evolved at the same rate as our culture; it was envisioned, created, and recorded by our ancestors long before we emerged from our "flat-Earth" *viewport.* Described and codified while we were deeply immersed in an *anthropomorphic* and *geocentric* mindset, the early pre-Christian records depict a God that is masculine, judgmental, controlling, revengeful, and punishing. This is the version of God that asked Abraham to sacrifice his son Isaac as a test; and later "gave" his favorite son, only to let him be crucified for "our" sins. Two thousand years later this primitive and primal vision of God still persists. During those two thousand years, many wars were initiated to eliminate, punish, or save those who believed in slightly different versions of this God. Five hundred years ago, this same God also "instructed" the leaders of the Christian church to torture, punish, and execute those who spoke of a new theory that placed the Sun, instead of the Earth, in the physical center of our solar system. From a three-dimensional worldview, this historical vision of God makes very little sense. However, from the perspective of an expansive Multiverse, this one vision of God actually makes a special kind of sense; it describes one more possible way that humans can express themselves in our universe of infinite possibilities.

GNOSTIC CHRISTIANITY

There are both ancient and modern Christian communities that have a very different interpretation of God, his message, and the meaning of *Christ*. Called *Gnostic Christianity*, this tradition predates even the formal church. The most important *gnostic* idea, the desire for a more direct and personal experiential spirituality, predates even Christ.

From its Greek and Latin roots, the word *gnostic* means knowing or knowledgeable. Ancient *gnostic* traditions often involved specific esoteric knowledge and practices, but the direct and personal spiritual experience has always been the common and critical identifier of *Gnostic Christianity.* Because of its focus on direct experience, *Gnostic Christianity* is a living and continuously evolving tradition.

Focused on their direct and personal religious experience, gnostic Christian authors see the living Christ inside each of us as the one who teaches. These writings often include common ideas such as we all equally belong to the oneness of creation, we are not our bodies, we exist eternally, and we are infinitely greater than we could ever imagine! This tradition also seems to fully allow for our most recent science.

Even though I was raised and confirmed within the Methodist faith, I was never convinced by the idea of the Christ that I saw portrayed in the texts and lessons of my religious education. His message and temperament seemed inconsistent, and his words, at least those taken from the King James Version of the Bible, often did not resonate within; to me they did not seem to describe a wisdom originating from a deeper place. This path simply did not ring true in my mind or my heart.

Many years later, I began learning about a different Christ, the Gnostic Christ, quoted and described in the scrolls found at Nag Hammadi. This version of Christ taught about a way of living, sharing, and being, which I found refreshing and distinctly different than the Christ of my Sunday school education. For the first time in my life, I began to feel a meaningful connection, a *resonance,* with this Christ.

The Nag Hammadi texts were recorded on papyrus during a period of time that has been identified as 40-200 years after Christ's death. Subsequently hidden in an urn for protection, they were finally discovered in 1945 along the upper Nile in Egypt. The scrolls were initially scattered by profiteers, but after many years of difficult tracking and focused acquisition, they have been reassembled into a nearly intact collection. Recently, scholars and researchers have been granted expanded access, and today high-tech equipment is busy reading, decoding, and analyzing the scrolls without having to unroll and possibly damage them. Worldwide, many researchers are hard at work making new translations universally available.

Written in the ancient Coptic language, these texts contain some of the earliest recorded recollections of Christ's teachings. The "Gospel of Thomas" is a text that parallels the four Biblical gospels (Matthew, Mark, Luke, and John), but it describes and brings to life a very different and much more mystical Jesus. Just as with the four gospels of the Bible, these passages from Thomas were not recorded directly by firsthand witness; they were, however, scribed at an earlier date than those from the Bible.

Not only was the Bible recorded later, but its passages have also been reinterpreted, re-edited, and rewritten many times over the generations to satisfy changing political, organizational, and social needs. Over time, the agenda-driven hand of man has dramatically re-

144

shaped the Bible, while the "Gospel of Thomas" stands unaltered. Because they were hidden for so long, the early Gnostic texts escaped this editing; they were never subject to the same political and social forces.

Through this relatively unaltered vista, a very different picture of Jesus emerges; he consistently reminds others to see his experience as an example. He explains to his audience that he is no different from any of them, and that everyone has within them the ability to experience and interact with life just as Christ does. He speaks about the way of serving God by directly serving others. He also teaches that no one else can function as an intermediary to God, because the true religious experience is always very direct and personal.

In the Book of Thomas, Jesus speaks like a guru, a shaman, or a man who understands how to navigate in a world where anything is possible because everything already exists and is waiting for each of us simply to become available within. He appears as one who long ago painted a picture of how to live freely in a quantum Multiverse.

Here are some translations from the Coptic Book of Thomas:

> "Whoever discovers the interpretation of these sayings will not taste death."

> "Those who seek should not stop seeking until they find. When they find, they will be disturbed. When they are disturbed, they will marvel, and will rule over all."

> "If your leaders say to you, 'Look, the (Father's) imperial rule is in the sky,' then the birds of the sky will precede you. If they say to you, 'It is in the sea,' then the fish will precede you. Rather, the (Father's) imperial rule is inside you and outside you. When you know yourselves, then you will be known, and you will understand that you are children of the living Father. But if you do not know yourselves, then you live in poverty, and you are the poverty."

> "Have you found the beginning, then, that you are looking for the end? You see, the end will be where the beginning is. Congratulations to the one who stands at the beginning: that one will know the end and will not taste death."

"Be passersby."

"If two make peace with each other in a single house, they will say to the mountain, 'Move from here!' and it will move."

"If they say to you, 'Where have you come from?' say to them, 'We have come from the light, from the place where the light came into being by itself, established [itself], and appeared in their image.' If they say to you, 'Is it you?' say, 'We are its children, and we are the chosen of the living Father.' If they ask you, 'What is the evidence of your Father in you?' say to them, 'It is motion and rest.'"

"His disciples said to him, 'When will the rest for the dead take place, and when will the new world come?' He said to them, 'What you are looking forward to has come, but you don't know it.' "

"His disciples said to him, 'Is circumcision useful or not?' He said to them, 'If it were useful, their father would produce children already circumcised from their mother. Rather, the true circumcision in spirit has become profitable in every respect.' "

"Jesus said, 'Look to the living one as long as you live, otherwise you might die and then try to see the living one, and you will be unable to see.' "

"For this reason I say, if one is (whole), one will be filled with light, but if one is divided, one will be filled with darkness."

"Whoever is near me is near the fire, and whoever is far from me is far from the (Father's) domain."

"Jesus said, 'Images are visible to people, but the light within them is hidden in the image of the Father's light. He will be disclosed, but his image is hidden by his light.'"

"You examine the face of heaven and earth, but you have not come to know the one who is in your presence, and you do not know how to examine the present moment."

"Whoever drinks from my mouth will become like me; I myself shall become that person, and the hidden things will be revealed to him."

"It will not come by watching for it. It will not be said, 'Look, here!' or 'Look, there!' Rather, the Father's kingdom is spread out upon the earth, and people don't see it."

Again, these ancient texts seem more like the words of a quantum physicist than of a religious teacher. This Christ reminds us that our being is much greater, deeper, and more full of possibilities than most of us realize. He teaches how to live in multi-dimensional and timeless space. He speaks of oneness in a holographic sense (within and without) as he reminds us that we all have the same abilities that he has and that everything we seek is already present within. ***The ancient Gospel of Thomas describes almost every idea that I discuss in this book.***

Contemporary Western Christian Gnosticism

Gnosticism is a form of Christian spirituality based upon our direct and personal experience with creation or God. When I read contemporary works of *Gnostic* Christian literature, three things immediately strike me. First, the philosophy, message, and attitude of the *gnostic* Christ, as expressed through these individual writers, are surprisingly consistent, but relatively simple, when compared to Christ's teachings from the King James or similar versions of the Bible. Secondly, they include much of the feeling, tone, ideas, and expression normally found in the Eastern spiritual philosophies. Often these contemporary *gnostic* writings seem almost mystical, portraying a vision of Christ that is similar to that of the Nag Hammadi texts. Thirdly, they seem to originate from a mindset that recognizes, accepts, and integrates a world built upon both *relativity* and *quantum physics*.

One of my favorite contemporary Christian writers is Paul Ferrini, who was originally led to his vision of a living Christ through the *Course in Miracles*. Much of his intuitive, thus revelatory, writing makes Christ sound like a cross between Buddha and David Bohm, the physicist. What follows are examples of this wisdom, which he attributes to his direct communication with the living Christ:

"The only prisons of this world are the ones of your own making!"

"Ultimately, the end of human suffering comes when you decide that you have suffered enough, when each of you, in your own lives begins to ask for a better way," and "No one way is better than another."

"All healing happens thus: as illusions are surrendered, truth appears. As separation is relinquished, the original unity emerges unchanged."

"This world is a birthing place for the emotional and mental body. Physical birth and death simply facilitate the development of a thinking-feeling consciousness that is responsible for its own creations."

"You have only one person to forgive in your journey and that is yourself. You are the judge. You are the jury. And you are the prisoner."

"As soon as there is the slightest perception of inequality between you and another person, you must understand that you have left your heart. You have abandoned the Truth."

"All experience happens for one purpose only: to expand your awareness. Any other meaning you see in your life experience is a meaning you made up."

These beautifully poetic descriptions of how to live in the *Multiverse* are fully built upon a Christian foundation. When viewed through an awareness that understands both Christianity and how the *Multiverse* works, they all make perfect sense.

JOSEPH CAMPBELL—THE POWER OF MYTH

A professor of mythology and comparative religion at Sarah Lawrence College, Joseph Campbell's lifelong quest was to understand and explain the information, message, and power of our mythology. Shortly before Campbell's death in 1987, Bill Moyers created the PBS television series *The Power of Myth*, which was based on Campbell's book of the same title. Through this series, Moyers introduced

Campbell's work to a large American audience and to the world at large. The series included extensive and absolutely wonderful discussions and interviews from his final years, and illuminated, for many of us, the deep and hidden meaning and extraordinary power of our mythology.

Like Gnostic Christianity, Campbell's ideas are based on his intuitive understanding, and they resonate fully with the fundamental *quantum* principles of the *Multiverse*. Joseph Campbell's deep love and warmth for humanity and his beautiful and poetic way of expressing his wisdom are unparalleled. Here is a small sampling of quotes from his books and lectures.

> "We must be willing to let go of the life we planned so as to have the life that is waiting for us."

> "If you do follow your bliss you put yourself on a kind of track that has been there all the while, waiting for you, and the life that you ought to be living is the one you are living. Follow your bliss and don't be afraid, and doors will open where you didn't know they were going to be."

> "Follow your bliss and the universe will open doors for you where there were only walls."

> "If you can see your path laid out in front of you step by step, you know it's not your path. Your own path you make with every step you take. That's why it's your path."

> "The cave you fear to enter holds the treasure you seek."

> "If you are falling...dive."

> "If the path before you is clear, you're probably on someone else's."

> "We're not on our journey to save the world but to save ourselves. But in doing that you save the world. The influence of a vital person vitalizes."

> "The first step to the knowledge of the wonder and mystery of life is the recognition of the monstrous nature of the earthly human realm as well as its glory, the realization that this is just

how it is and that it cannot and will not be changed. Those who think they know how the universe could have been had they created it, without pain, without sorrow, without time, without death, are unfit for illumination."

"The goal of life is to make your heartbeat match the beat of the universe, to match your nature with Nature."

"Gods suppressed become devils, and often it is these devils whom we first encounter when we turn inward."

"Find a place inside where there's joy, and the joy will burn out the pain."

"The big question is whether you are going to be able to say a hearty yes to your adventure."

"The experience of eternity right here and now is the function of life. Heaven is not the place to have the experience; here is the place to have the experience."

"We save the world by being alive ourselves."

"All the gods, all the heavens, all the hells, are within you."

THE "NEW AGE"

I feel particularly blessed to have grown up in America in the 1950s and 1960s. For most of my life, there has been a stable but rapidly expanding economy which helped many of us to meet, and even exceed, our biological and physical needs. With this new affluence, we had unprecedented free time, since far fewer hours were spent procuring food and shelter. This free time allowed many of us to delve deeply into our spiritual side and experiment with a wide variety of alternative ways of living on this Earth. I know of no other time or place in our planet's history that has created this level of opportunity for so many.

My generation took enthusiastic advantage of this unique luxury, expressing our individualism and spirit in a multitude of varied ways: rock and roll, drugs, sexual exploration, political and environmental activism, social activism, alternative religion. We explored and experimented with lifestyles, meditation, Eastern philosophy, *quantum physics*, new healing methods, television, computers, Internet, world travel, and even space travel, as we expanded our understanding of what was possible in a multitude of ways that our parent's generation could never have imagined. As time passed and people started writing about their experiences, a new culture was built around these new ideas. With the Internet unfolding in perfect timing, the various new and reworked ideas quickly spread to other people and cultures around our world. Simultaneously, these new methods and technologies allowed us to discover that these "others" from completely different backgrounds were having similar experiences. We are now in the midst of a true, worldwide renaissance; new inspirational and inter-connecting *information* is expressed, shared, and incorporated as never before.

As with any emerging, natural, and unregulated human movement, it is being shaped by a wide variety of influences. Some ideas and new technologies are the result of sincere spiritual explorations, while many others may only be the result of self-serving or monetarily driven agendas. Ultimately, it makes little difference because all these forces are working together to form the eclectic body of material and *information* that is grouped under the broad, but unofficial, banner of "New Age."

While much of this new material may be far-fetched, lacking thoughtful examination or requiring a disproportionate amount of faith, we will also find powerful examinations of a deeper, inter-connecting truth, along with many new ideas about different ways of approaching and improving our lives. The common ground that drives this movement is a collective recognition that our old ways, ideas, and methods no longer fully serve humankind. "New Age" philosophy, while clearly a "mixed bag," also provides many of the revolutionary ideas that are helping to guide the emergent paradigm shift. In many ways the emergence of New Age philosophy parallels the emergence of our new physics.

By this point, readers may also realize that certain ideas in this book are very different from much of contemporary, New-Age thought. The principles of this book and "New-Age" philosophy certainly diverge in a number of significant ways, and I will discuss these differences in context.

"New Age" Ideas Derive from "New Thought"

Isolated references to many of these "New Age" ideas can be found in the recorded philosophy of small, scattered, groups from our past. Like blips on a radar screen, the moments of history that allowed for the exploration of these ideas were brief; they came and went. These ideas seemed radical and generally failed to attract larger audiences until the early 1800s, when a talented and well-publicized group of alternative writers and philosophers coalesced under the banner of "New Thought." This "New Thought" philosophy promoted many of the ideas that we associate today with "New Age" thinking.

Originating in New England, the "New Thought" movement was influenced and guided by several well-known literary figures, like Emerson and Thoreau. Evolving from German Transcendentalism, its basic unifying principles were that God or "Universal Intelligence" is supreme, eternal, and universal, but at the same time, divinity dwells in each and every person. They explained that we are here to love one another and our actual thoughts determine the manifestation of our lives. Dedicated to healing the general human condition, "New Thought" philosophy was widely published and spread rapidly.

The earliest identifiable proponent of what became known as "New Thought" was Phineas Parkhurst Quimby (1802–1866), an American philosopher, watchmaker, healer, and inventor. He became known for developing the healing system that he called "Mind-Cure"; a system based on the ideas that illness originated in the mind because of our belief system and that a mind open to God could cure any illness.

God and the Four Gospels were always a major part of Quimby's original "Mind-Cure" philosophy, and this Christian core idea continued as the greater "New Thought" movement took hold. Quimby himself had worked closely with Mary Baker Eddy, who went on to form the *Church of Christian Science,* a contemporary church whose philosophy also derived from the "New Thought/Mind-Cure" movement.

An important principle of "New Thought" was an idea that lately has been renamed the "Law of Attraction." Today, this concept is enjoying a renewed popularity, largely due to a recent movie, *The Secret.* The idea is that we all attract those very people and things that are like us—like attracts like. Other original and important "New Thought" principles such as "positive thinking," "creative visualization," and "conscious-languaging" are also being revisited.

As mentioned, Ralph Waldo Emerson was a major influence on this "New Thought" movement. Although Emerson wrote on many topics, his thematic core was consistent. In his writings, he felt and communicated the deeper interconnection and unity amongst all things. A few samples from his work follow:

"All separate things are made of one hidden thing."

"The world globes itself in a drop of dew."

"The heart and soul of all men being one, this bitterness of his and mine ceases."

"Christ is in all."

"Every heart vibrates to that iron string."

"I am my brother and my brother is me."

"God appears in a myriad of disguises."

"We are all it, and so is the leaf, the stone, the grass, the mountain, and the cloud."

The next major reemergence of this philosophy was when Joel S. Goldsmith sifted through and filtered many of the principles of "New Thought" to create his "Infinite Way" program in 1947. *The Course of Miracles*, reportedly channeled in 1965 and published in 1976 by Columbia Medical University psychologist Helen Schucman, followed soon after. While not directly derivative of "New Thought," it is clearly an elaboration and an expansion of many of these same principles. Recently, popular writers such as Joy Brugh, Wayne Dyer, Marianne Williamson, Deepak Chopra, and others have reintroduced these same principles to the mainstream population through books, lectures, and workshops.

The "New Age" movement has never been as formal, coherent, or organized as "New Thought" because it has no parent organization, established doctrine, or set of standard practices—it is genuinely "grass roots" in its origins. Many who practice this contemporary version have minimized or even abandoned formal religious ties; they often favor the direct *gnostic* experience. Notable exceptions to this secularization and lack of formal organization are several new, alternative Christian practices, such as the *Unity* church and the previously mentioned *Course in Miracles,* both of which are quite organized and thriving.

Visualization, Conscious Language and Positive Thinking

Today, we have unparalleled access to a plethora of self-help books, websites and workshops based upon positive thinking, visualization, manifestation, *conscious languaging,* and similar techniques, all borrowed from the "New Thought" movement. Extremely popular in the West, these methods often seem to work, sometimes quite well; they work because they address the hidden aspects of a deeper truth. Because they also appeal to our individual, but ubiquitous, desire to control more of our lives, these techniques can also cause us great difficulty. They are only tools and, like all tools, (knife, axe, or auto) they can create problems if not used correctly.

Positive thinking is almost self-explanatory. The general idea is that holding positive ideas in our minds manifests more positive in our lives. **At the most profound level, "positive thinking" is the conscious, continuous and grateful recognition of the enormous beauty in life and the purposeful interconnection of all things. This kind of positive thinking has the potential to dramatically alter the appearance of our individual worlds in many wonderful ways. If someone wants to choose one practical change to make in his or her life, I would personally recommend keeping a joyful, open attitude. This is often the single most helpful change any person can make to redirect her or his life.**

Conscious languaging is the intentional practice of becoming fully aware of the intended and hidden meanings of the actual words we use every day. In this practice, there is a conscious focus on constructing our spoken word to more accurately reflect our desired intentions. Since everything is a form of *energy,* this includes the words we speak and the thoughts that formed them. Words have a hidden energetic power; therefore, we have much better results when we use them carefully and thoughtfully. Our words are our everyday prayers.

We have all, feeling exasperated, exclaimed things such as *"I can't do this!" "I'm really dumb!"* or *"I don't have enough money."* This unconscious habit is extremely self-limiting; energetically, these words act at levels we cannot understand. Such negating word choices, are, at the very least, reinforcing the exact opposite of what we want. Instead, if we turn these thoughts into more positive expressions, our words become our allies in setting our intentions and achieving our goals. For example, instead of saying, *"I can't do this,"* we could say, *"I am learning to do this!"* As we observe and analyze our actual word choices, we are often shocked to discover just how many times every day we verbally thwart ourselves with this unconscious sabotage.

Visualization and *manifestation* are closely related "New Age" techniques. In both, we create and hold a "vision" of something that we want to change or acquire in our lives. Instead of just "wishing" for this change in one's mind, this practice usually involves imagining or experiencing the change as if it has been already completed, which deepens the experience. For example, we might visualize our ideal future family life. Not limited to just our visual sense alone, many forms of this practice will also holistically incorporate aspects of sound, feelings, smell, taste, memories, and thoughts.

Some variants of this method focus on our feeling the future—we are not trying to visualize or force a specific outcome; rather we are attempting to allow the future to reveal itself to us through our feelings. The difference here is very significant. With this method, we are allowing and feeling rather than trying to create and control; our *feelings* are much closer to our core *resonant vibrations* than our thinking is. By focusing on *feeling* and allowing the future to present itself, instead of doing, creating or manipulating, the activity is no longer centered just in our thinking brains; instead we get to explore a deeper terrain that lies much closer to the core of our *being*.

Since linear "time" does not exist in the *Multiverse,* "feeling the future" does not require knowing things that have not yet happened; every possibility already exists somewhere in the Web. **The present moment informs the "past" and the "future" equally through waves of information that travel in all directions.** Our relationship to any possible "future" is similar to our relationship to our "past." Future memories are not that different from past memories, and one day we will be able to access both in much the same way. **This also means that our actions in the present can influence the past; this is a**

radical, revolutionary, and life-changing realization. I explore this extremely important idea more deeply in *The Architecture of Freedom.*

If we try using any of these techniques to influence or control future outcomes, then these attempts are destined to become personal explorations of what occurs when we become too *attached* to specific outcomes. The techniques become exercises that provide wonderful opportunities to learn about how we are prone to create *attachments* and what this really means. As a laboratory to learn about *attachment* and the true "creative" process, this type of experiment, if approached consciously, can serve us very well. By fully exploring our will through various forms of creative manifestation, we can come to better understand our *beingness* and how our *Multiverse* really works. If what we really want is a new Mercedes, then some of these self-help methods can help make that goal a reality. However, one day we will realize that we are still driving that once brand-new car to the same old enervating job; at some point, that Mercedes will no longer satisfy us.

One life-lesson that this exploration will likely teach us is that our attempts at control will very quickly *dampen* much of the inherent joy, magic, and surprise that naturally accompanies our living in the flow. Our deepest and most *resonant* creations can only emerge when the river of life is flowing fully.

Because of our deep *identification* with the physical *illusion*, we often completely miss the real potential of these "New Age" techniques. Typically, we lack the sensitivity and perspective to know what is "best" for us; our personal material wishes are rarely more than unguided attempts to satisfy our egos. Of course, any of our personal journeys spawned by such attempts will contribute to our soul's experience and inevitable expansion. All types of exploration are wonderful in the bigger picture; ultimately, any process that explores life's possibilities will lead to the growth of our individual and collective soul.

Because of some very common misunderstandings about these techniques, they often result in the formation of more *attachments.* There is always a very real possibility that the participating individual becomes *attached* to their specific imagined outcome; this *attachment*, then, introduces all the issues and inevitable suffering that are commonly associated with all types of *attachment*.

As they are taught in the West, these techniques often involve focusing most of our energy upon the level of our material world. In my experience, it is the rare contemporary Western workshop or teacher that encourages students to investigate the deeper levels of *Being*; we are a culture that has grown accustomed to quick and easy gratification. ***Manifesting such predetermined and personal egoic visions, while completely possible, will often require us to swim against the natural currents that flow within the river of our lives.***

There is, of course, nothing "wrong" with playing with these methods, especially if we consciously recognize what we are doing and understand that these methods are just tools for exploration. Through this type of play, we will likely discover that we are extremely powerful beings and really can "manifest" the very things that we desire—but they often come with unexpected, unseen costs. Along with this discovery, we also learn a great deal about our individual *desires* and the nature of *desire* in general.

We all benefit from more play in our lives, and exploring these techniques provides us with an interesting and focused curriculum for this play. However, to create real growth through permanent change, we must ultimately shift the core *resonant vibration* of our *being*. While a *positive attitude* and the practice of *conscious visualization* can move us in that direction, lasting change requires that we explore a territory that lies deeper within. Eventually our core beliefs, our personal fears, our embodied physical and psychological traumas, and our culturally inherited attitudes must be illuminated so they can be "seen," explored and released; only then will we be able to freely travel far from our old vision of the universe.

We Create Our Own Reality

One popular "New Age" belief that leads to a great amount of confusion, disappointment, and disillusionment is the deeply empowering idea that "we create our own reality." While this statement correctly expresses a very fundamental truth about our *being,* a better understanding of this principle is necessary if we are to avoid confusion. The issue is that many people interpret this statement to mean that simply thinking or holding a thought long enough or strongly enough will cause it to manifest. They usually then discover, to their disappointment, that this level of application does little to change the appearance of the outside world, especially in the specific ways they had hoped for. As a result, many give up exploring this interesting idea, only to drop back into the old and familiar "there is nothing I can do about it" attitude.

What they are missing is the critical effect and importance of our deeply held *beliefs* and the body's memories. The vision will only have the ability to manifest if it is in *vibrational resonance* with the deepest levels of our *being.* Without this *resonance*, the vision remains constrained by our old beliefs about the limited three-dimensional nature of the *Multiverse.* Because we are so conditioned to look for quick and easy fixes, this very important doorway to a deeper truth often is overlooked or ignored.

A more effective method for "creating" our own reality begins with our understanding that in the timeless *Multiverse,* every reality already exists. We do not ever "create" any reality; therefore, it is more accurate to say, "we find an already existing reality," "we open the path to it," or even, "it finds us." The path and reality already exist, but we can only find them through relaxing and allowing. The not-so-subtle distinction between these two ways of viewing our place in creation illuminates a key to new awareness that is necessary for our journey towards freedom.

The most important ingredient for change is our willingness to open ourselves to and be grateful for life no matter how it presents itself to us. The path towards freedom requires that we master being joyful "observers" instead of trying to be the "doers" or "controllers." Jesus said, "Be a passerby." As we learn to trust and let go of the habitual need to control our lives, our individual journeys appear to become smoother and, at the same time, inexplicably amazing in their serendipitous unfolding. For each of us, our individual natural flow is always perfect; often the difficult part of the process is learning how to relax and enjoy the ride. What we eventually discover is that when we are being "passersby," we are not so much observers as we are participants, for there is ultimately nothing that exists outside of us. We become one with the ride.

Law of Attraction

The popular "New Age" idea called the "Law of Attraction" (sometimes expressed as the concept of "like attracts like") is also a borrowed principle from the earlier "New Thought" movement. This "law" states that if we are of a particular mind set, or vibration, we will attract into our life those things that support us in our journey. This "law" is actually describing *sympathetic resonance*.

Much of the popular interest in the "law of attraction" is due to its promise of attracting personal and material things, such as money, luxuries, fame, success, and romantic love. Instead of observing what arrives in our lives and developing gratitude for the flow of life and *being*, this practice is often used to help attract a specific list of personally desired things. Since the list of desires is usually the product of our egoic drive to feel safe, special, successful, or even superior to others, using it this way ultimately only creates more separation. These personal wants or needs rarely originate from the deeper desire to become more open to our *being,* or truth. This narrow focus completely misses the amazing potential of this principle.

Practiced this way, the technique can still "attract" the desired results, but any initial success will probably lead to new and unexpected problems or, put more positively, "learning opportunities." Again, there is nothing "wrong" with exploring our personal desires this way because every explored pathway leads to more experience, the raw material for all growth and evolution. What we eventually learn is that the fulfillment of our personal wish list, by itself, will not result in sustainable joy, inner peace, or personal freedom.

The "Law of Attraction" is a misunderstood reframing of a deep-level truth about the *vibratory* nature of our *Multiverse*; we can easily understand this principle through the classical physics of *sympathetic vibration*. Things of similar *vibrational resonance* do appear in our lives, but not because we "attract" them to us; more accurately, our evolving *vibrational resonance* allows us to interact with new parts of the universe that were previously hidden from us. These things attract us to them because we have become open to and available for these new experiences. ***When the "Law of Attraction" appears to work, it is because our universe shifts as it responds to our own changing vibration. From our old perspective, it may appear that we attracted these changes, but the deeper truth is that we opened to allow ourselves to take a journey to a different part of the Multiverse—one that now resonates with who we really are. In one sense, we shifted to a parallel universe where these desired results***

were ready to manifest. So we did not attract the results. Instead, their vibration attracted us!

A more accurate description of the "law of attraction" can be expressed as *that which vibrates as you vibrate will appear to manifest in your life.* A more descriptive name would be *the law of sympathetic resonance.* Many of these "New Age" spiritual techniques have their foundations set in this deeper truth. By exploring any of these techniques consciously, *information* from the depths will rise and alter the way we understand our lives.

"Be Here Now"

For good reason, "Living in the now" is a concept that is embraced universally throughout the "New Age" movement. *Be Here Now*, the 1971 book by *Ram Dass,* discusses aligning ourselves with this fundamental operating principle of the Web. Once we learn how to recognize, trust, feel, and fully enter the "now moment," we will have uncovered our most direct, personal, and time-tested portal to freedom. This simple method, available to everyone, allows us to experience our lives in a completely different way.

When adults are introduced to this idea, the first difficulty they encounter is simply to understand what these words actually mean. Before the influence of cultural imprinting, young children understand this "now" moment naturally; in fact, it is the only way of *being* for newborn infants. As an adult, relearning this way of experiencing requires a focused and consistent practice. In our busy and demanding culture, it is difficult to remember the meaning and feeling of *"being here now."*

Once we re-learn this natural way of experiencing life, there remains the constant hurdle of navigating and participating in our culture as we open to *"being in the now."* Our demanding culture is always pulling us into some preplanned "future," while simultaneously reminding us of our "past." Just finding the *now moment* becomes a very difficult task. To stay in the *now* while functioning in our society, an adult must hold a clear and steady vision, one that is supported by a deep trust. It typically takes significant life experience, and therefore time, to build this vision and trust. Living in the "now" requires a solid foundation, deep commitment, and constant reevaluation, but once we get the hang of it, we discover that there is a completely different and amazing new way of moving through life.

It is a popular trend in "New Age" writings to refer to *fear* and *Love* as polar opposites. I believe that this comes from a misunderstanding of both *fear* and *Love* for, while they are certainly related, they originate in entirely different realms.

Love is the undisturbed life-force vibration. *Love* simply "is"; it is not a human conceptual creation. What we call "romantic love" is very different from *Love*. Romantic love is a concept formed within our three-dimensional realm by our egos. Romantic attraction may or may not involve the *resonance* that engages the deeper form of *Love*. Whenever and wherever we *resonate* within our depths, we experience more of this ever-present form of *Love*. This is a never-ending pool of *Love* which we can then share with all others; when we dive in fully, our typical feeling of separation entirely disappears. Within this deep pool of connection there is only *Love*: "One Love."

Fear is a three-dimensional idea that exists specifically for our *viewport*. Created by our egoic minds as a self-protective mechanism, it tends to work overtime to dominate and interfere with our natural awareness of the deeper type of *Love*. *Fear* is a completely natural byproduct of our physical individualization, but if left unexamined, it can grow to intensify the feeling of isolation that is inevitable within our exploration of physical separation. *Fear* can expand in our minds to fully block the recognition of our deep interconnection with all others. It is absolutely true that when our *fear* increases we feel less *Love, and* this is why these two appear to be opposites. It is impossible to be in *fear* and in *love* at the same time. Fear blocks our flow and our natural connection with *Self*. **Fear does have its three-dimensional opposite, and this opposite is freedom. When we are free, we experience our direct connection to this constant, palpable, and extremely powerful, deeper form of Love.**

Our deepest encounters with *Love* often happen spontaneously. They occur most often when we forget our egoic selves while caring for another, about another, or when our ego becomes "lost" through some joyful activity like music, dance, or love-making. The flow of *Love* may be experienced as a still moment that includes the satisfying presence of deeper truth; an awareness of our connection to everything, or a knowing that there is life beyond this physical realm. When this awareness is present, it means that we have dropped our fears, allowing us to connect to that which lies beyond our realm—at least temporarily. This is why we often experience *Love* while relating to another person; we feel *Love* when we place our full attention outside of our separate selves. *Love* is always available whenever our *ego,* and the *fear* that *ego* generates, are out of the way.

It is also helpful to note that we are never "in fear" or "in love," but rather we are always "in motion," moving in the direction of either separation or oneness. We might be moving towards separation, increasing our *fear* and impeding the flow of *vibrational Love,* or we might be moving towards oneness and experiencing a more open flow and a greater *resonance* with *Love.* As with all natural motion, it is normal for our connection with *Love* to shift cyclically within a natural rhythm built upon ebb and flow, like the tides. This cyclic change is not something we need to analyze and correct; rather it is simply something to observe. As always, *"be a passerby."*

Other Common "New Age" Misconceptions

Today, a popular "New-Age" belief is that one day, as human consciousness evolves, our Earth will reach a *critical mass* of evolved souls, which will spontaneously change everyone else and thrust our planet into a new, enlightened age. Some refer to this as the "Hundredth Monkey" phenomenon (based on the misinterpretation of a WWII-era experiment using monkeys on a small Japanese island— see *The Architecture of Freedom*).

Life provides the experiences necessary for the evolution of human consciousness; this is our purpose. However, the appearance of our lives is a direct reflection of our current state of individual consciousness. Therefore, as our individual consciousness changes, our personal view of the world changes accordingly. The rest of the planet does not need to evolve for our world to change; instead, as we individually evolve, we actually shift to a different universe—the one that best reflects our present, inner vibrational state. As we evolve individually, we actually then "appear" in an infinite series of parallel universes that each perfectly reflect the evolving state of our individual consciousness.

We do not ever have to change anything about the outside world; we only need to change the world that resides inside us, which provides the appearance of an entirely new reflection for the world outside. We are making these changes constantly, every time a new thought sinks in and affects our vibrational core. Our dualistic world is never going to be "fixed," for it is always changing to be the perfect reflection of our inner vibrational patterns.

Another popular view of our evolutionary process sees a "battle between good and evil," hopefully to be won by the forces of good. This also cannot occur because duality always requires these opposites—they must co-exist together. One extreme can never eliminate the other, for without this type of polarity, the physical world would not even exist.

Seeing the world as a "struggle" or "battle" is an individual choice or experiment that results in a particular flow of outcomes. The participant's world will reflect this mindset, one focused on the "struggle," and this belief will shape the same harsh viewport over and over again. This particular "experiment" is one that has been repeated many times throughout history with very similar results.

GEOMETRY OF DIVINITY (GOD)

"I am the Alpha and the Omega, and I am the Living One!"

The *Web of Possibilities* provides a structural foundation to explain the reasons for many of the mysteries within our lives, including many qualities we assign to God. There have always been births along with deaths, storms and blue skies, miraculous healing and sickness, good and bad fortune. Throughout our history these have been attributed to the hand of God; a God has often been assigned responsibility for those natural processes and connections that were too difficult for us to understand. Some of these processes can be explained fully within classical physics, but because many originate beyond the veil of our three-dimensional universe, much of this mystery can only be understood through the hidden architecture of the *Multiverse*.

For example, the idea that God is omnipresent seems quite impossible from a three-dimensional conceptual mindset. By understanding the intimate interconnectedness of all parts of the *Multiverse through enfoldment, parallel universes, holographic information storage,* and *timelessness,* we can understand the physics of how "God" could possibly operate everywhere at once. There is no need for an outside deity to pay attention to everything, remember all the details, and process all the *information,* because the *Multiverse* does all this by itself; the capacity for full "awareness" is built right into its structure. Since everything in the *Multiverse* connects to everything else, there are never unexpected surprises, just infinite natural connections allowing cause-and-effect to be generated at levels far beyond imaginations built upon our three-dimensional understanding. The *Multiverse* automatically behaves like a single, giant brain—God's brain.

Throughout modern history, notable scientists and philosophers have punctuated major scientific breakthroughs by declaring that science is finally on the brink of explaining God. Stephen Hawking stated that he believes our new *cosmology* will soon be able to explain the major remaining mystery attributed to God: life-force. A number of well-known computer scientists have announced that God will be found somewhere along the trail while developing *Artificial Intelligence.* Historically, with each major scientific breakthrough, we add to our understanding about the materials and energies that make up the physical world and gain insights into additional realms that lie beyond; but we still find ourselves with little or no understanding about the actual "force" that animates life. This force has never been identified or located, yet it is clear that without it, there is no life. **While the Web is revealed as an amazing playground for the dance of life, life-force itself remains the great mystery.**

SIN AND HEALING

"Sin," an important religious concept, has a very different meaning at the level of the *Multiverse*. Sin is defined universally as a violation of divine law and it is often described as an act that harms others. Because, in the Multiverse, there are no others, at this level, sin can be understood as an act or thought that is not aligned with our "own" best self-interest. Sin is the natural result of not knowing, trusting, or following our deeper truth. When witnessed from the deeper levels of *Being*, even the most heinous acts against "others" (and, therefore, ourselves) can be understood as little more than an "inefficient" waste of *energy* due to unnecessary or "wrong" turns while exploring the infinite possibilities.

Everything that we label as "sin" is actually little more than an unconscious or inefficient expenditure of *energy* when viewed from this deeper and timeless perspective. "Sins" are not even "wrong" turns because, ultimately, every turn that is possible must be explored and experienced for *Being* to become fully compassionate. It is the nature of life to fill every void and explore every crack and opening, and a critical part of our very purpose is to help *Being* do this, without judgment. We occasionally become lost in the exploration of our separation, but even this type of misadventure adds to our breadth of experience. Since everything is intimately connected, everything is also completely self-correcting because any natural system will always return to equilibrium over "time." There are no sins, mistakes, or wrong turns within creation; *Being* is witnessing, learning, and growing more compassionate with each and every experience.

We "sin" because we fear, or do not fully trust or understand the flow of *Being*. When we "sin," on some level it only means that we are not realizing that this act against "others" is ultimately an act against ourselves. The sinner is unaware that, at our deepest level, we are only a single *being* and there really are no "others" to even "sin" against. One reason that we universally feel guilt, shame, and pain in association with our "sins" is because, at some level, we intuitively understand that we have ultimately created more suffering and difficulty for our own *Self*.

What happens after we "sin" is often the bigger issue. We reflexively attempt to mitigate the intensity of any pain arising from this self-effacing choice by subconsciously restricting our energy flow or *life-force*. Emotional pain is largely the product of inner tension and resistance to the natural flow of *life-force*. When *life-force* is restricted or shut down, the body actually does experience less pain, but there is an enormous price to pay for this temporary relief. The body becomes less sensitive and less receptive in every way, which means it will not feel as much of anything. Unfortunately, this also means that this restricted system has become less *vibrant* and less capable of feeling *joy*. In the bigger picture, this is a form of self-defeating behavior that makes no logical sense. However, because of our unconscious and reflexive reactions to fear and pain, this behavior remains our most common response.

The most direct way to expand the flow of joy is to learn how to release our chronic tension. The first step in this process is exploring, discovering, knowing, and then embracing the deep sources of our tension. The ultimate source of all tension is the *fear* generated through our cultural sense of separation—the *fear* of existing and then dying, alone and isolated. The healing begins with a deepening of our recognition that we are always fully connected with everything in creation. This means that the path to wholeness, and ultimately freedom, must involve dropping all the judgment that separates us, and simultaneously embracing everything within—both the "sin" and the "sinner." All aspects of all "others" and ourselves can then be understood and seen as valuable, important, and equal parts of creation. ***Our personal development of this nonjudgmental embrace of all life is often the most difficult step in the entire process of becoming free.***

GROUP THINK

Group experiences are often enjoyable because of the tangible feeling of increased or *amplified* empowerment generated through *sympathetic resonance*. Something profound and exciting happens when "many act as one" and, because of this, political leaders and other powerful interests have always attempted to harness this powerful *resonance*. Governments, large corporations, and other powerful entities have historically exploited our human need to connect by attempting to manipulate groups of people in order to propel their own agendas. A study of world history reveals that organized religion has often been used as a tool to consolidate and manipulate large masses of individuals for other political or social purposes. In addition, the powerful agents of fear and nationalism have been incorporated into this mix to further *amplify* it. Fear, religion, and nationalism often are packaged together to divert otherwise peace-loving individuals into supporting war plans or similar large scale adventures.

This entire process involves exploiting natural *resonance* to unconsciously influence people. This type of mass manipulation would not be possible if more of us were aware and conscious of the source and meaning of this deep and powerful phenomenon.

Why Workshops Work

Many of us have had amazing experiences in group workshops. A special sense of empowerment occurs in these focused group settings, whether the group consists of two or two thousand. Often we later discover that these same techniques do not seem to work as well or as consistently once we are outside of the group workshop environment. There are two primary reasons for this perceived loss of effectiveness.

Firstly, the center of our physical body is located close to our heart, far from our brain. Many years ago, a very wise friend reminded me that my brain, located in my head, is an "extremity," and that I needed to understand my brain and my "thinking" in this context. Recent experiments in neurophysiology indicate that there are far more nerve passageways sending information from the heart to the brain than there are from the brain to the heart[3]. This neural arrangement could be interpreted to imply that, biologically, the brain is designed to be subservient to the heart. Our hearts are clearly much more than just pumps for our blood.

We already know this, for we commonly include the phrase "heart" to describe things that are near and dear to us. We intuitively understand that real internal change must originate in our hearts. Of course, no single part of our *being* works alone; all parts of our *energy* field must be involved and "on-board" for creating meaningful change to our *vibrational* state. However, it is only the heart which can be always found at the center of our deepest internal and interpersonal interactions.

In workshops, we usually interact very closely with other people. The group engages our natural connections and we automatically move away from the busy, egoic nature of our individual minds. The workshop environment helps shift the center of our expression away from our brain and closer to our heart, where our experience of "others" is much more fluid and interconnected. The group process works largely because our hearts naturally become interconnected as we work and spend time together.

The second factor is that the *energy* of "others" influences every one of us. Workshops, retreats, and other group exercises can create

---

[3] (Rollin McCraty—Heart Math Institute, multiple studies)

significant *energy* fields that surround and support us. As mentioned, focused interactions allow many individuals to become "tuned together" in *resonant harmony*. When we are thinking and acting together as a unified body we discover a group *resonance* which is often *coherent,* cohesive, and *reinforcing.* This is not unlike a sports team being "in the zone" together, or the "electric" experience of an extraordinary rock concert.

Most of us have had the opportunity to sit in a crowded arena during an exciting ball game, surrounded by fans who are intensely devoted to a team. Do we really care about the game that much, or are we there for the powerful, focused group experience? If our chosen team wins, there is an enormous amount of excited *group-vibration*; it often means the team moves on in the playoffs and all the fans will get another chance to have this connected and empowering feeling. Political rallies depend on this *coherence* and its *reinforcement;* Hitler was a master at manipulating this type of group dynamic. For many, the real purpose of attending large group events is not for the game, the concert, or the political message; it is more about the human desire for deep, personal interconnection. *Coherent, reinforced,* and *resonant vibratory* group experiences often fuel excitement, energy, and ecstasy.

Resonant group experiences can be tremendously powerful and transformational for the individuals involved. Focus groups and workshops can create a deeper awareness of both our deep interconnectedness and life's amazing possibilities. Many types of spiritual groups and workshops have excellent records assisting individuals through drug rehabilitation or emotional recovery. Support from those who have shared a similar experience allows us to more easily release and move beyond our personal fears and habits. Turning our attention towards others, whether it is a guru, lover, family member, or friend, helps us to step outside of our reflexive, egocentric focus. Then, with a greater sense of interconnection and fewer feelings of separation, we naturally experience more joy and ecstasy.

Unfortunately, maintaining this high level of interpersonal *resonance* becomes much more difficult once we go back to our daily lives. The massive inertia of our existing culture, and the divergent interests and demands of those with whom we interact daily, will always have their own strong influence upon us. Outside of the intentionally focused group environment, our culture's powerful and often chaotic influences make it more difficult to have these *coherent vibratory* experiences.

Within our realm, we all participate in the great group experience called life. There will always be aspects of each of us that are sensitive to and interact with those influences that originate outside of our separate physical bodies; for example, the *energy* field of a longtime spouse is usually a very strong influence on our system. Our boss, coworkers, children, friends, and favorite newscasters all strongly interact with our individual *energy field*. Again, we are reminded that at the deepest levels, just as *quantum physics* describes, everything is intimately connected. Everything that exists, ever existed, or could exist is directly connected throughout the entire *Multiverse*. At the deepest levels of our *being* there is no such thing as an outside influence because there is, after all, only one thing in creation.

Cults

During difficult times, we may benefit from outside help and guidance. We might seek new information or a person who can help us find a path that better meets our needs, or at least one that is relatively free from the suffering and pain that we have been experiencing. From time to time charismatic individuals or new ideas appear, offering a practice, theory, or a way of living, which in that moment holds some special promise. With a strong charismatic leader or concept that resonates for a particular time and situation, a community of like-minded thinkers may gather. This community will then *reinforce* and empower that particular message by providing additional structure for physical, *vibrational,* and emotional support and comfort. Many powerful, self-reinforcing, *resonant* communities have formed this way.

Sometimes the devotees of these groups are encouraged to focus most of their energy and attention upon their leader. This type of absolute devotion—called *Bhakti yoga* in the *Hindu* tradition—is a very powerful method for evoking change through a process based upon *ego surrender*. As in any devoted relationship, over time the devotees begin to *resonate sympathetically* with their leader or guru. An entire group's attention upon a single, common focus can shape and form a powerful and *coherent group resonance*. Whether it is at a rock concert, political rally, sporting event, workshop, or within a cult, this type of group experience is real, powerful, and very tangible. The power of Hitler's filmed political rallies is still fully palpable today, even though the films are only available in grainy black and white with poor sound quality. Many books have been written about this "group-think" phenomenon.

What turns this type of group experience into a cult is the intention to link the devotee's experience directly to the leader or the philosophy. **Cults, by definition, do not guide devotees towards personal freedom; instead, devotees are kept dependent in a form of prison, with walls formed by their beliefs about the leader or philosophy.** Along with a prescribed set of rules and punishments, there is often an unrealized promise of "freedom," but only at some future time. Most cults make it extremely difficult for individuals to leave.

Because of the strong *coherent resonance* within these guru-centered group environments, devotees naturally feel a deeper level of comfort, interconnection, and empowerment. With this type of connection, the common feelings of separation and loneliness can be more easily forgotten. At some point in their surrender process, individuals usually experience a major release of tension, resulting in powerful, ecstatic feelings.

It would be easy and natural to associate these ecstatic feelings directly with the physical form of the guru or leader and this is sometimes encouraged. However, by making this association, the devotee has missed the critical connection between their new feelings of ecstasy and the specific act of quieting their *egos*. The deeper, hidden truth is that when we surrender our personal desires and focus on something outside of these, ecstasy will follow—no guru is necessary. Ecstasy appears because, by focusing attention on "others," our *egos* experience less isolation and separation. This is the simple secret behind the power of "group experience." *The best teachers are those that guide their students towards this realization and thus to their own personal freedom.*

We all have the innate ability to experience ecstasy through conscious *ego-surrender* without the involvement of a guru or group. This requires a high level of self-trust, often gained from a lifetime of conscious personal work and self-examination. Achieved this way, the joy and ecstasy are fully sustainable because there is no dependence on another's behavior or form. Along with this joy and ecstasy will come an expanded experience, with the full range of all other human feelings and emotions and a fresh opportunity to understand life in a different way.

Groups built around charismatic leaders are usually not very flexible or durable, since they are based on a particular physical, but temporary, form—that of the guru. The *resonance* of the system will usually fall apart when the group structure shifts or erodes due to a change within the leadership. If the charismatic leader leaves the group or dies, then *resonant feedback* stops. Those groups dependent on the physical presence of their guru never survive this type of transition. Any parts that do happen to survive often only mimic and distort the original message: they continue to exist in name only. The current forms of many modern religious organizations are the direct result of this type of de-evolution.

After the loss of their leader, devotees frequently enter a period of shock; it can be very difficult for the newly and often suddenly dissociated individuals to navigate their own paths to the previously experienced joy, release, and ecstasy. Trained to be dependent on their guru, they may have mostly forgotten their own innate and direct connection to the actual source of their ecstasy: *Being* itself.

Ironically, it is often when their guru becomes emotionally or intellectually compromised that adherents have the most difficulty in freeing themselves. The followers of Charles Manson and Jim Jones followed behaviors that few of them would have believed in or supported on their own.

Religious Group Experience

The core principles of most world religions point towards a deeper and common truth. Unfortunately, over time this deeper awareness has been lost or buried because of the agendas, business practices, and politics of many modern religious organizations. Most of our more established religions rely upon an imposed hierarchical structure that intentionally separates the individual from the directness of their spiritual or religious experience. These established organizations all rely on specific designated individuals—priests, preachers, rabbis, imams, or similar trained individuals—to interpret religious experience for the common practitioner. By relying on these surrogates, the practitioners systematically become disempowered. Completely forgetting about their direct connection to the deeper source, they no longer experience the richness or understand the deeper meanings of the original spiritual message. By design, these organizations condition individuals to believe that they do not have direct personal access to authentic religious experience.

I recently noticed this phenomenon played out at a memorial service. While the preacher, who had not known the deceased, was leading the service, the words and sentiment were very impersonal. Most of the guests were fidgeting and not very interested or emotionally involved. Due to the indirect and impersonal nature of the experience, his words did not evoke the beautiful presence of the person whose life we were honoring. Toward the end of the memorial service, family and friends had a chance to describe their personal experiences and feelings. The preacher was astounded as this short, scheduled part spontaneously evolved and expanded into its own full-length service, becoming the most memorable part of the ceremony. We shed tears, recounted stories, shared laughs, and sang songs as the person we were honoring suddenly came to life in all our hearts. Direct and personal experiences manifested this powerful connection to the spirit of the person whose life we were celebrating.

HEALING AND CHRONIC TENSION

The Western medical model traditionally only treats deep energetic blockages at the superficial level of physical symptoms. We have become dependent upon pain-relievers and mood-altering drugs, either as prescriptions or as self-medication. We resort to surgery to deaden nerves, relieve pain, loosen muscles, and realign joints; all conditions created only because the body has shifted out of its natural balance and harmony. Addressing our issues at this level just temporarily masks this deeper imbalance. Medications can interfere with important natural biological processes and often produce serious, problem-compounding side effects. Surgery creates scar tissue that may eventually lead to new physical blockages and additional problems.

The only real and permanent solutions to chronic health problems involve lifestyle, diet, and addressing the causative psychological or emotional issues at their source. As improbable as it seems from our Western mindset, the source of chronic medical conditions or disease can usually be reduced to basic and fundamental imbalances in our lifestyles, often involving our *attachments*. At their deepest level, all of our food and diet issues involve *attachments*.

Psychiatry and psychology incorporate many useful tools to help address physical tension or emotional suffering; unfortunately, the most recent medical trend has been for doctors to prescribe more psychoactive drugs instead of using psychotherapy. For severe and chronic cases, the pharmacological approach can work as an effective tool for immediate intervention. However, it has become clear that over time, the use of these drugs can result in many undesired and often dire side effects. Once the life-threatening phase of the emergency has passed, psychological help and counseling become the most valuable tools for the healing process. To facilitate a more permanent healing these established medical practices need to be integrated with newer *holistic* techniques that address the entire body and greater *Being*.

Our bodies, like everything in the *Multiverse,* are *holographic* and store a complete record of our emotional and psychological journey. If encouraged to communicate, our bodies will direct us to the hidden keys for identifying and releasing any chronic tension that impedes our natural flow. While these blockages are still minor, they happily signal their presence with small aches and pains; at this point in the process, they are gently "asking" for further attention. Only much later, if ignored or left untreated, do these early warnings evolve into the diseases and debilitating conditions that require more extensive intervention and treatment.

One of the most effective methods that I know of for addressing the deep-level tension at its source involves the integration of several forms of bodywork. Initially we might use the bodywork for temporary pain and stress relief to create the time needed to address the deeper issues. Later, bodywork from a well-trained and skilled therapist can set the stage for release and insights into the deeper *attachment* issues that originally stimulated the tension. A talented therapist cannot only locate and work on the physical areas of tension and damage, but they might also be able to suggest correlations, insights or possible causes. At the very least, once we learn where we are holding tension, we can begin to pay attention to these areas, and eventually they will provide internal clues about the problem's source. Through skilled bodywork, we can learn about the deeper causes of our tension while we also receive needed, short-term relief.

Along with insightful bodywork, it is also usually beneficial to include physical therapy and alternative body-centered techniques such as ecstatic dance or yoga. In addition, numerous new forms of psychological therapy such as Emotional Freedom Technique (EFT) can improve the effectiveness of psychological counseling.

If approached with an open mind, all of these techniques can work together, synergistically, to help us heal our minds and bodies; and through this process, we will discover more about our expansive selves. Our primary goal is always to enhance the free movement of *energy* and *information* (flow), so that we can live to our maximum potential. If we are prepared for and open to this type of growth, and if use these tools effectively, we will begin to identify, and then better understand, our *attachments* to things within this *illusion*. The healing section of *The Architecture of Freedom* contains a more in-depth discussion of these and other methods.

HEALTHY DE-ATTACHMENT

Healthy "de-*attachment*" is an extremely difficult idea for the culturally trained Western mind to understand and embrace. We typically only understand detachment as "dropping out," often through desensitizing ourselves with drugs or some form of busy activity. This "shutting down" inevitably moves us even further away from the interconnection and vibrancy that we actually desire. Many of us have personally experienced this type of detachment, and we have certainly been close witnesses to this behavior through others. We all understand that this type of detachment is never healthy in the long-term, even though it often involves culturally approved methods, such as prescribed anxiety medications, alcohol use, excessive exercise, or work addiction.

This is a detachment from life itself and because it usually leads to isolation and desensitization, it is almost the polar opposite of spiritual *de-attachment*. When Westerners first are introduced to the Eastern idea of *attachment,* we reflexively prejudice this new concept with our old cultural understanding of detachment. Because of our preconceptions, we tend to judge *de-attachment* as a destructive and antisocial practice.

In the healthy spiritual form, the goal is to de-attach, not from life, sensitivity and experience, rather from the relentless grip of the ego and the drama it continuously creates. We accomplish this by learning how not to identify with the things that appear in our

illusionary expression of life, including our bodies and our egos. To understand and benefit from this positive form of de-attachment, we must first incorporate the deep "knowing" that we are not these bodies; also, we are not what we do, own, or create in this life.

According to Eckhart Tolle in his first book, *The Power of Now*, "When our consciousness frees itself from its identification with physical and mental forms, it becomes what some call pure or enlightened consciousness or simply 'presence'."

We cannot just decide to become *de-attached.* This way of being evolves over time and is a natural byproduct of a deep, personal *inquiry,* focused upon understanding just who and what we really are. Individuals who have mastered the healthy form of de-*attachment* still live full lives in this world—actually, they are living fuller lives because they feel more as they experience more of the deep, direct interconnections. They also do not become disturbed when things do not go as planned or desired, and they have an inner understanding of the balance in all things. For this same reason, the *unattached* individual does not become excessively giddy when things appear to be going particularly well. They still make plans but remain flexible, knowing that plans will evolve and change in each and every flowing moment. They understand what the American mythologist Joseph Campbell was expressing when he said, "If the path before you is clear, you're probably on someone else's."

This way of moving through life leads to an even-tempered *presence*, a sense of being a "passer-by," even while fully participating in life. Along with participating in more of life while enjoying all of its flavors, practitioners experience an expansion of joy and a reduction of stress. Once a person is able to *un-attach*, they realize that their involvement can be more like watching a good movie or playing a video game; life is just revealing one more way for the fully interactive movie-like *projection* to play itself out. While it takes on the qualities of a more impersonal adventure, life also becomes much more fun, enjoyable and exciting. With an understanding of *attachment* and a mastery of de-*attachment*, we free ourselves from the endless repetition of being stuck in our own, self-manufactured quicksand. Instead, we flow freely down the "river of life," always fully engaged through our direct and intimate connection to the ever-evolving *present moment*. Our lives become more *flowing* and *interconnected.*

"OTHERS" WHO ARE FAR LESS FORTUNATE

This brings us directly to an extremely difficult and challenging conceptual hurdle involving the very real and profound human suffering observed throughout our world. This book's philosophy may be interesting as an abstract concept, but how does it serve those in dire need? It might be easy for someone with resources and a healthy body to talk abstractly about spiritual healing, but what about the less fortunate person who was born in a war-torn, impoverished country or is crippled with a lifelong disease? What does this philosophy mean for all the innocent children around the world who were born into horrible conditions?

The answer to these kinds of questions is multilayered and requires a much deeper understanding of who we really are and how we inhabit the *Multiverse*. First, if we recognize our interconnections, eventually we will understand that we are actually all only one *being.* At this point, we will also realize that at some level we possess a single, collective soul. This was what Jesus was referring to when he commanded, "Thou shalt love thy neighbor as thyself. There is no other commandment greater than this," and when he said, "As you did it to one of the least of these my brothers, you did it to me."

From this deeper perspective there are no "others," so this unfortunate *being,* who is living with the less than ideal body or in a terrible situation, can be understood as an expression of some dissociated or rejected part of our greater collective *Self.* As each of us opens and becomes more sensitive to these deep interconnections, we begin to experience much less separation. When we fully recognize that this "others" life is a part of our own experience, our entire way of living and relating to these "others" shifts. Ripples from this new *awareness* can then transform the appearance of the lives of everyone, including those who are "less fortunate." As our relationship to the world, which is presently understood to be "out there," shifts within, then, because we are all completely interconnected, there will also be less separation, isolation and suffering on the "outside."

This suffering of "others" is also a valuable teacher guiding our always-evolving, collective soul. A part of our collective *being* is having this difficult experience so that we may all have another opportunity to witness, gradually understand, evolve, and become more aware and compassionate. Through this and similar experiences, our collective soul deepens and increases its understanding about our interconnection and our *attachment* to that which is *impermanent*. As we unravel our true relationship to these "others," we will naturally begin to love and care for them, and incorporate these isolated and rejected parts of our own *being*. These separated parts and isolated individuals can only become healed and whole through each of us learning how to live our lives with the recognition of the oneness of all of *Being*.

The outside world often seems harsh from the perspective of our temporary individual separation, but it always seeks balance, completion and perfection through all of its natural processes. We easily can understand how storms, earthquakes, and disease are natural or necessary parts of a balanced ecosystem. In a similar way, the existence of these extremely challenged individuals serves the ecosystem of our consciousness. All parts of our collective awareness are reflecting exactly what is needed in this and every moment to create balance and deepen our *awareness*. Since we are all intimately interconnected, we ultimately all benefit and evolve together. Life is cruel and cold only if one is deeply *attached* to the *illusion* of time, our bodies, and this particular physical manifestation of life.

Our most challenging and difficult experiences produce the greatest opportunities for growth. This applies to both the individual and the collective. Steven Hawking has expressed deep gratefulness for his extremely reduced physical condition. This famous cosmologist fully credits his disease, ALS, for his focus, achievements, and even his satisfaction with life. The unique focus that we gain through extreme hardship can often become a rich catalyst for an amazing amount of growth and productivity. However, sometimes deep within the most perfect opportunities for growth and opening, it still becomes necessary to cast off our physical bodies because they no longer serve the deepest needs of our soul. As humans, we need to remind ourselves repeatedly that we are so much more than these bodies.

HEAVEN AND HELL IN A QUANTUM UNIVERSE

As we live our lives, we naturally and unconsciously move throughout the Web according to a path being charted by our always-shifting *vibrational resonance*. We might experience that which we encounter in any moment as pleasant or particularly difficult. However, no matter what type experience we are having, it is always possible to instantly leap to an entirely new type of interaction. The nature of multi-dimensional space and *enfoldment* allows for all possible places and experiences to always be directly adjacent and, therefore, fully accessible to all of us in each and every moment.

Dramatic shifts to completely different *parallel worlds* or *universes* are fully possible while still embodied in our three-dimensional form, but it is more common to make smaller and more gradual shifts. Our cultural concepts and ideas form a semi-rigid structure that effectively acts like an anchor or shock absorber, limiting or *damping* the maximum degree of movement or change that is *probable* during any one adjustment. Along with this *damping*, the human form generally responds better to more gradual shifts that include longer periods for readjustment and acclimation. Therefore, most shifts within the *Web* tend to be gradual and subtle; we often make our most dramatic adjustments only at times of trauma or deep personal work.

We also make a particularly large and unique shift at the time of our physical death, when the conceptual anchors and embodied memories of the body cease to be a factor. Upon physical death, our natural state of *vibrational resonance* is liberated from its many physical, conceptual, cultural and emotional restraints. Freed from the constraints of our old *viewport,* our individual *being* then migrates automatically to those areas of the Web where it is more naturally *resonant.* Dramatic movement is possible at this special time because our souls are much less restrained by our old ideas and patterns. Released from the constraints of body, our spirit spontaneously shifts to that very part of the Web that is in complete *harmony* with our soul's core *vibration*—the deep-level *resonance* of the *self.*

This is the real mechanics of the various spiritual teachings about "heaven or hell" after death. Our *core vibration* at our time of death fully determines what our existence will look like in the "next" expression of our *being.* There are no rewards, judgments or scorecards—this is just a natural and organic process based upon *vibratory resonance.* This means that everything we do and pursue in our lives (and the resulting deep-level changes), all shape our *core-resonant-being.* It is our *resonance* at that very special moment that then determines the *vibratory* environment for the "future" expressions of our life.

In the *Multiverse,* there is no such thing as a reward for "good" behavior, or punishment for "bad" behavior, because all manifestations and expressions are always only the direct result of cause and effect. Instead of rewards, there are simply additional opportunities to meet with the *unexamined* through new events in our lives that help us understand more about the wide variety of different influences shaping the human condition. With an open mind, even the most challenging and difficult human experiences can be experienced as expansive, beautiful and rewarding. Evolution is ultimately about becoming more: more open, more experienced, more aware, more compassionate, and more integrated. We grow bigger and include more of everything as we become more of who we really are. This also is the path to the deepest and most enduring experience of Love.

It is essential to remember that these concepts are being described using our old ideas about time. We now understand that "time" does not unfold in a linear fashion within the *Multiverse*. At the deeper levels, all "lifetimes" exist simultaneously and are interwoven into and inform the ever-present "now" moment.

EXPLORE IT ALL

We live in such a wonderful and interesting time and place. A defining characteristic of our time is the unprecedented and almost universal access and availability of information, methods and teachers. Today, workshops, self-help books, therapies, contemporary religions, and spiritual practices can be found almost everywhere. Our generation has easy access to an amazong wealth of opportunities, all designed to help individuals lead happier and healthier lives.

At the same time, this outside exploration is fully optional because immersion into family, friends, or career is an equally valuable approach. When we are open and actively engaged in any aspect of our lives, we are participating fully. As John Lennon sang, "Life is what happens to you when you're busy making other plans."

Many of the terms and practices I discuss in this book are borrowed from common "New Age" methods and technologies; I am, after all, a child of these times. Once we begin to see the bigger picture of how and why these things work, and sometimes do not work, we can then focus on the practices that serve us best, refining them to create our own individualized practice. Regardless of which specific paths we explore, ultimately the journey itself serves us. If we approach our spiritual journey with an open energetic curiosity, then joy and self-empowerment will naturally follow.

Over the millennia, as we collectively evolved, we formed and solidified a series of consensus beliefs about how our world and the greater universe works. ***This set of assumptions, understandings, values, concepts, and practices forms our cultural paradigm. Our inherited paradigm fully determines what we think is possible and how we live our lives.*** That which we call "common sense" is always dependent upon our worldview. Today, for the most part, the vast majority of humans are connected through a general agreement with only a few small subcultures embracing a significantly different view. We all have many personal experiences that fit and support this collective and deeply entrenched *paradigm,* and this constant reinforcement causes our common vision of the universe to seem extremely solid and real.

However, simultaneously, each of us also has a much less-discussed collection of personal experiences that do not easily fit into this collective vision. We all manage these unusual experiences in different ways. Sometimes we may not even recognize the parts of episodes that fall outside of our cultural understanding. ***Most of the time we are not even consciously aware of the aspects of our experience that lie outside of our normal expectations because we all have a filter built into our brains that psychologists have named cognitive dissonance.*** Our conscious minds are only capable of processing new information if it fits with things that we already understand and believe: our known worldview. ***This means that things or experiences that fall outside of our worldview are not even recognized, seen, cognized, or processed. We build our concepts from the bricks and mortar of older concepts that our minds already understand.*** Since it has evolved for efficiency and survival, this automatic psychological filter restricts the amount and variety of our conscious input. *Cognitive dissonance* is our "conscious awareness *damper.*" To help our minds to work efficiently in our three-dimensional universe, we automatically cast aside strange or unexpected ideas, so they are not even processed by our conscious awareness. Due to *cognitive dissonance*, we often "miss" the more unusual, "out-of-the-box" aspects of our daily experiences.

Sometimes we may notice unusual aspects but not understand them, so we consciously choose to remain quiet. If what we notice actually disturbs our sense of well-being, we might then forget or bury the incident for our own self-protection. One common way we do this is by keeping our minds engaged in more comprehensible activities like socializing, working, or playing.

Occasionally, some of us recognize our "out-of-the-box" occurrences as the actual "paradigm busters" that they really are, but still consciously choose to ignore them because of various personal concerns. We might find ourselves embarrassed or uncomfortable discussing these unusual encounters; within our current culture, there is no acceptable context for processing these events. Many of us attend churches, or join spiritual or support groups in our attempts to normalize, discover the meaning of, or provide a language for these "out-of-the-box" experiences.

However, not all unusual encounters are lost in "the prison of our paradigm." Some will break through the veil to be expressed as visual art, books, poetry, song, movies, insights, spiritual experiences, and personal growth. In the next section I discuss specific "paradigm-busting" experiences—some from my own life and some from others'.

SECTION THREE—DIRECT EXPERIENCE

INTRODUCTION

There are times in all of our lives when the veil that encloses and defines our three-dimensional world becomes a little more transparent, revealing hints of extraordinary things beyond. Through nature's design, we are afforded only brief glimpses into this more expansive geometry and reality. Sometimes, when we have a particularly stressful or life-threatening event, this boundary softens spontaneously. At other times, our sightings into the bigger and more interconnected existence occur during our most relaxed or open states. We all have these openings—they are an important part of life. As we mature, we start to accumulate experiences that hint at, or even reveal, the existence of a much-larger universe. Some of these experiences hint at a much different connection between the body and *awareness* than that which we usually expect or understand; some glimpses help us realize that "time," with its relentless rate of marching and its single direction, might not be so absolute. Others point to a different type of intimate relationship with the world and other beings. In one way or another, all of these "out-of-the-box" stories illustrate that there is much more to existence than just our familiar world.

The experience might involve a powerful interconnection, or an instant understanding or insight, that cannot easily be explained. We might notice an uncanny premonition about a friend, a sensation of déjà vu, a clear knowing of how the next scene will unfold, a clear sense of being outside of our body, or an encounter that seems completely outside of normal time. We use words like "understanding," "premonition," "hits," and "feelings" to help describe this type of awareness, but it often arrives first as just a very subtle feeling.

Our Western culture does not generally recognize that we have this type of direct access to a more expansive existence; it provides little or no support when we have these types of experiences. Many parts of our culture do not even recognize that this level of interconnection is possible, even though most people can recall at least a few of these expansive moments.

Growing up, I frequently noticed when my personal experiences did not conform or make sense within our prevailing paradigm, and I was fortunate to experience a dramatic one of these life-changing events at just the right age. I was old enough to understand the implications, but young and innocent enough not to be boxed into the prevailing cultural vision. Over time, as I became more familiar and comfortable with my own "out-of-the-box" experiences, I found myself able to discuss these experiences with others. What I quickly discovered was that I was far from alone. Most "others" were also having these unusual encounters and experiencing the same struggle about sharing them, often keeping these experiences to themselves because of concerns about how they might be perceived by others, especially in professional circles.

What follow are stories describing a few of my direct encounters—those that had the most dramatic influence on my life. Afterwards, I relate similar stories of friends and family before I conclude with "out-of-the-box" experiences that we all share. I now recognize that these non-conforming events have been occurring constantly throughout my life. When I stop and observe carefully, I notice that they actually happen many times every day. I have grown so comfortable with these "paradigm-busting" occurrences that they are now a normal part of my life and seem quite ordinary. Most of these stories describe my first glimpses—those early impressionistic experiences that left permanent marks on my psyche.

MY PERSONAL EXPERIENCES

TIME SLOWS IN AUTO ACCIDENT

I place this particular story first because it was the crucial life event that marked the actual beginning of my conscious exploration of the edges of our awareness. It was not the first time something like this had happened to me, but it was the first time I felt the strong need to reevaluate my inherited worldview.

I was seventeen and driving my parents' car to a friend's house, which was about twenty miles away, when this unexpected experience completely changed the way I understand and relate to "time." My family lived in a rural area, one where the long stretches of empty, two-lane highway invited young male drivers to push the limits of speed. I was, as I often did in those days of feeling "bulletproof," traveling over 100 miles per hour with an empty road ahead. On a long straightaway, another car inexplicably pulled out right in front of me. Rather than slow or brake, I swerved into the clear, oncoming lane. Completely unaware of my car, the other vehicle again changed lanes; they never saw me—a car traveling over 100 MPH was probably unexpected and outside of their *viewport*! With no place left to go, I swerved off the road into a very rutty apple orchard filled with rows of large and closely spaced trees.

At over 100 miles per hour, all of this happened very fast, but in that "moment," I found that suddenly I had all the time in the world. My sense of movement and "time" was instantly altered; my awareness became a slow-motion slideshow revealing one clear frame at a time. Each framed view held as long as I needed it, and then it automatically shifted to the next frozen scene. Each scene was a perfectly arranged, still view of exactly what I needed to see for the next maneuver, and I was always given the perfect amount of "time" to look at everything. I was aware of every direction, every corner of the car, and every tree. From my perspective, *time* actually stopped and then progressed "one frame at a time" in slow motion; I had all the *time* I needed to decide which direction I should steer and how to go about each maneuver. I slowly and deliberately executed the movements required to negotiate each single still frame before looking again, and I repeated this process with each slowly advancing frame. It was as if I had a remote control and my "slideshow" was clicking ahead six inches at a time, pausing just long enough for me to observe, react, and then assess the next six inches of necessary movement. I cleared dozens of trees by inches and then, a few hundred very bumpy yards later, I brought the car to a halt in a cloud of dust. Once fully stopped, I could finally look at my overall situation, and for the first time, as I returned to my normal consciousness, I noticed that my heart was pounding wildly. After a long pause to catch my breath, I assessed that not only was I alive and well, but I had not even visibly damaged my parents' car (although I am quite sure that its suspension was never the same).

At that moment, a sudden awareness overwhelmed my *being*. *"What in the world had just happened?"* I turned around and there I was, sitting in the middle of a potato field, a thicket of big trees behind me, and a clear set of crooked tire ruts marking the maneuvers that I had just completed at over 100 mph. *"This really happened!"* In that moment, my life was forever changed. I had encountered something that my training and upbringing could not explain, and I was suddenly thrust into a completely new perception of time and reality!

Recently I was talking about "time" perception with my sister and she recounted a similar story. She was involved in a multi-car accident on the freeway and for her, "time" also slowed to almost a stop. She had "time" to observe the rapidly changing situation and knew exactly when and how to duck to avoid flying objects and glass.

Many athletes have similar stories of "time" slowing, and being in a "zone" while playing their sport. Motocross racers will talk about the "time" they had to make conscious decisions while flying through the air, mid-accident. In my musical career I have had similar experiences. When playing a fast and complex musical phrase, I occasionally have all the "time" necessary for feeling, phrasing, and executing the quickly passing phrase.

Similar "time" altering experiences occur in many situations throughout all of our lives. Our personal sense about the passing of "time" is extremely malleable and not nearly as fixed or rigid as our timepieces, schedules, and culture would have us believe.

OUT-OF-BODY EXPERIENCE

We normally assume that our individual awareness always is tightly connected with our physical body, but this does not have to be the case. In special circumstances, our center of awareness can actually move away from our physical body—sometimes far away. We might suddenly witness our awareness focused above our body, behind it, off in the next room, or moving about freely somewhere far away and unrecognizable to us. Many of us can remember having this type of experience at least once in our lives.

According to doctors, patients, and many others, this is a common phenomenon, especially when great stress or traumatic injuries are involved. Patients involved in accidents or emergency surgery have sometimes discovered that their conscious *presence* was fully aware, watching, hearing, and understanding the EMS workers or doctors who were working on their otherwise unconscious bodies. They

185

relate how they were suddenly watching from the tree above their wrecked car, the corner of the operating room, or from some other remote place as others attended to their damaged bodies. In this remote, but conscious, state of awareness they sometimes had life-changing spiritual experiences, including instantaneous reviews of their entire lives. Occasionally, they can even recall having made the thoroughly analyzed and conscious decision to return to their original, but very compromised, body.

This type of occurrence is so common it has been named the *near death experience* or *NDE.* Many of these stories have been collected and recorded by patients, doctors, researchers, and psychologists; scores of books examine this interesting phenomenon in great detail. Of course, these types of memories can only be told by those who have experienced this traumatic separation from their bodies, but later were able to return to their bodies and continue with their lives; we hear nothing from those who made the decision not to return.

Quite spontaneously, and in a very dramatic way, this also happened to me. My first experience of my awareness leaving my physical body was extremely emotionally disruptive; it left me feeling unsettled for several years. In my late twenties and newly married, I was living in Lake Placid, New York. While fully enjoying the circumstances of my own personal life, I was also deeply disturbed by the general condition of the world. The widespread poverty, wars, human injustices, environmental problems, and our inherent cultural shortsightedness all disturbed me deeply. At this time in my life, I had been meditating daily for almost ten years, and it was rather automatic for me to leave behind my thoughts and the day-to-day world while practicing. Our apartment featured a picture window that directly faced the beautiful, snowcapped peak of Whiteface Mountain, which sat majestically, just twenty miles away. Every morning I would meditate for half of an hour, using this beautiful mountaintop as a visual *mantra* to help focus my attention and clear my mind.

On this one particular day, instead of dropping into my usual relaxed meditative state, I felt overcome by a sudden powerful bodily "whooshing" sensation, and then quickly found myself rapidly moving through a tunnel-like space, as if flying. In the next instant, I was seeing and experiencing the world from the very top of that mountain, looking directly back to the town of Lake Placid, where my actual body was still sitting. I had no awareness of a physical body or any sense of being cold, but my conscious presence and visual sense were fully centered right there on that mountaintop. After a few moments, I had the clear and strong "knowing" that I had an important and critical decision to make; I could continue on this journey through this exciting new space or return to my body. I also somehow "knew" in that instant that, while this state of being was tempting to explore, if I continued further, I would lose the will and ability to return to my new wife and the life that I had been living. I panicked, lost my focus, and my consciousness was instantly pulled into a similar, but darker and more rapidly spinning tunnel. I then found myself sitting in my living room, but now shaking uncontrollably and completely covered with sweat.

I know of many others who also have, for various reasons, found their conscious awareness temporarily shifted to somewhere other than within their own body. When my brother was a teenager, he often found himself observing his own back while walking down the road. Another friend, as a pre-teen, repeatedly floated outside her house and was able to watch her sleeping body through a window. My wife once hovered above herself watching her own conversation. While the center of our awareness is normally associated with our bodies, it also seems that, in special conditions, it has the ability to roam free of this physical boundary.

After I experienced sitting firmly on a mountaintop that was more than twenty miles from my body, I stopped my deep-meditation practice for a number of years, afraid that I might have a similar experience and not have the ability, or more accurately, the desire to return. Upon reflection, it is now clear that during that period of my life I was fundamentally unhappy because of my perception about the problematic nature of our human existence. Back then, my connection to my body and this life was not always joyful, and some large part of me desired escape. I also sensed that it was not my time to make this transition; I still had much more to do and learn while in this body. The path I chose produced two beautiful children, an exciting career, many great adventures, many dear friends, a sense of humor about the human condition, and a strong appreciation for all of life and its amazing mystery. I am now certain that our body and our conscious awareness are two different things; they are interrelated and interconnected, but still separate. *We are not our bodies. This is one of the most important realizations on the road to true freedom. This one clear realization serves to liberate us in many wonderful and diverse ways.*

HEARING THE VOICE OF "GOD"

The most unexpected encounter of my life occurred on the deck of a sailboat in the middle of the Pacific Ocean, while living in Fiji working with the Peace Corps. One beautiful evening, I visited a friend's boat. Lying on the deck watching the full moon, the stars, and the night sky, I noticed that the spinning anemometer at the top of the mast was nearly the same size and shape as the round, full moon, which was also right overhead. My fascination with shapes and form took hold, and only a small shift in my position was necessary to place the anemometer directly in front of the full moon. It was a perfect fit—the wind meter covered the entire moon. The spinning wheel unexpectedly created a powerful strobe effect and I instantly became disoriented; I felt like I was being pulled up a tube of spinning light while my entire body felt like it was twisting and undulating. When I "landed" after a few long moments, the spinning stopped and I heard a deep booming voice that I instantly "understood" to be "the voice of God." This voice was as real, deep, resonant, and vocal as any physical voice that I have ever heard, but I later learned that no one else on board heard anything. The voice commanded my full attention and I quickly lost all sense of my body and everything else that was around me. This booming voice was warm, laughing tenderly, and seemed extremely friendly as it carefully and clearly spoke the words: *"Tim…You always worry about everything. Just remember that there are many different ways of looking at anything. Everything has positive and negative aspects or interpretations. Find the positive in every situation, focus on these positive aspects and your entire life will change."* Suddenly it was over; the voice was gone, and I was, once again, lying there on the deck with the clear night sky, moon and stars.

I have no explanation for what happened that moonlit night, except that the spinning wheel centered on the moon somehow facilitated a change in mental states, not unlike classical hypnosis. The incident made a deep impression and I immediately began searching for ways to change how I viewed and approached life. It took me many years and many additional reinforcing insights to fully understand this message and its meaning, but today this awareness forms the foundation of my life philosophy. ***That message was distilled, directed, and worded so that I could hear it with my mindset that night. It was essentially relaying one of the most fundamental principles of this book: If we change our attitude, we will change our entire world. Every possibility exists, so the world that we*** *perceive is always, and only, a reflection of our most deeply-held beliefs.*

VOICES IN THE ROMAN FORUM

One year after my encounter with that voice on the sailboat, I was visiting Rome for the first time. I was not an architect in those days, or even thinking about a career in architecture, yet that morning I found myself powerfully drawn to the ancient ruins of the Roman Forum. It was before sunrise, but due to jet lag I had already been awake for a couple of hours, and, just as the new day broke, I found myself completely alone in the Roman Forum. The light and the peaceful feeling that morning were exquisite, and I found myself meandering around the forum in a joyful trance. I started to hear faint murmurs and assumed merchants or others had arrived to hawk their wares, but when I searched to find the source of the voices, there was no one to be found. These voices became louder and I noticed that they projected the power and cadence of public orators, but I was unable to understand the words. I began to feel the hairs on my arms rise and tingle as I realized that what I was hearing sounded like ancient Roman oratory of the type that had once been the daily custom in this very place. This multilayered symphony went on for twenty minutes or so, and involved dozens of different voices from different building locations within the ruins. Gradually, as other people entered the area and the Forum began to become active with the bustle of morning life, these mysterious voices disappeared.

To this day, whenever I think about that early morning many years ago, I always feel that same powerful, tingling sensation up and down my arms and legs. Once again, my strong sense of a marching linear time was torn asunder as *information* from a bygone era suddenly became available and alive in the present.

MEET NEW CLIENT AT POOL

One day about twenty-five years ago, I was at my usual neighborhood pool for lap swimming when I engaged in a conversation about building with a friend, who happened to be a contractor. A complete stranger, a woman, walked up and asked, "I couldn't help overhearing your conversation. Is either one of you an architect? I need one very badly."

Later I found out that her journey that day had been every mother's nightmare involving a car full of screaming children, traffic, broken plans, and repeated disappointments, as she drove them between several beaches and pools, looking for a place to swim. All her usual and alternate swimming spots had been closed for repairs or other reasons so, with a car full of screaming kids and the temperature well over one hundred degrees, she was at her limit. After her numerous failed attempts, she eventually stumbled upon this unfamiliar pool, only because she had again been rerouted by road construction. Later she told me that she immediately had sensed that this was the pool where she was to be that day, even though she did not know why.

This chance meeting led directly to several of my more interesting architectural jobs, a significant change in my life direction, and one of the most significant friendships of my life. This chance meeting altered the trajectory of my life in ways that I could never have imagined in those first moments. It has taken me many years to appreciate the way life seems to have a rhythm for bringing to us the exact experiences we require. The pattern of our interconnections often seems to lie far outside any possible logic or planning that our brains might rationally construct.

OTHER UNEXPECTED MEETINGS

Returning from a three-month stay in Greece in the mid-1980s, I found that I had a short layover in London so I thought I might try to look up a dear friend. I did not have his address or phone number with me, but I knew that he was studying at the London School of Economics, so I got off at the Tube stop closest to his school. I understood that finding him at night in this crowded city without any address or phone number was a long shot at best; my hope was that the school would have a public directory. Concerned that it was quickly becoming too late to even search for a directory, I started running as I exited the subway door. Moving too fast for the crowded conditions, I rounded a blind corner into a dense crowd and immediately collided with someone coming from the opposite direction. The collision was so violent that we both tumbled to the pavement, bags and bodies fully intertwined; not before or since have I accidentally bumped into someone hard enough to cause such a fall. During the process of untangling our backpacks and briefcases, our faces met and, to our mutual great surprise, there I was, fully entangled with the very friend that I was in London to visit.

Astounded, we shared a long, deep, and joyful belly laugh that christened another wonderful evening together.

Another time, years earlier, I was on an extended adventure hitchhiking around Canada and the United States. At various backpacker gathering points, such as state park campgrounds and hostels, I had repeatedly crossed paths with one particular person, and noting that these unplanned meetings were frequent and continued to happen, we naturally became friends. One day we stepped out of the spontaneous character of our relationship by planning to meet in Yosemite on a particular day several weeks in the future, but that meeting did not happen as planned, so I just continued with my journey. Two months later, while camping on a long beach that runs down the west coast of Vancouver Island in Canada, I suddenly had a strong desire to have salmon for dinner. I hitched the twenty miles to the tiny fishing village of Tofino, only to discover, to my disappointment, that there was just a single fishing-boat in port at that time of day. As I walked on board, I heard voices from below, so I shouted down into the hold. Up came my friend with an enormous salmon in her arms. She saw me and said, *"Looks like we are having salmon for dinner."* As it turned out, she was camping within 100 yards of my camp along an open and endless beach that stretched for miles and miles. She was also a professional gourmet cook, and that night I enjoyed the best-tasting fire-grilled salmon that I can remember.

As I was proofing this very section of the book, a very interesting event unfolded. While not a formative experience like the others, this story is included because of its particularly unusual timing. A year earlier, my wife and I had a life-threatening misadventure on a small boat in the waters of the Fiji Islands. Four other strangers were also on that boat, and, fortunately, we all survived. Exactly one year later, my wife and I were visiting San Francisco. As we were riding the ferry from Sausalito to the city on a very windy day, we saw a boat almost capsize so we began to recount our adventure from the previous year. For some unknown reason this was the first time that we had ever reviewed this near disaster. Five minutes later the ferry docked and we walked directly to the adjacent cable-car stop. Immediately a voice behind us shouted, "Hey remember me?" We turned to discover one of the other four people who were on that boat in Fiji the previous year; he lived in Prague and was in San Francisco for only that one day. The timing of our discussion, this immediate chance meeting, and my editing, once again, seemed to completely defy the bounds of normal probability.

Over the years, I have had many other chance meetings that seemed to defy normal probability, but these first two were the most life-changing. We are connected in many ways that are invisible to us, and all of these connections directly affect our lives. These improbable occurrences continue in my life, but now I view them as quite normal, and sometimes, even expected.

WARNING IN THE NIGHT

I was camping with a friend on a beach directly south of Big Sur, California in the early 1970s. It was very dark when we set camp, and being inexperienced and unfamiliar with local conditions, we inadvertently set up too close to the base of the large cliffs that define the seashore of that region. At some time in the dark, early-morning hours I awoke, heart racing, full of terror, and quickly woke my friend to insist that we move. I had to admit to him that I did not know why— I just knew we had to move. Deep in his sleep mode, he did not want to be bothered, but because of my persistence, we pulled our sleeping bags out towards the middle of the beach to a spot where I felt much more comfortable. Before we even climbed back into our bags there was an enormous rumble and several large boulders fell right into the area where we had been sleeping. It was not until the full morning light that we fully recognized the critical importance of our last-minute decision to move. In hindsight, some part of me obviously understood that our original sleeping spot was poorly chosen; but just what was it that woke me from the deep sleep, and how did all that unfold with just the right timing? Some protective part of my *being* must have been connected to *information* that was not a part of my normal waking consciousness and sense of linear time.

FLOOD OF MEMORIES

About fourteen years ago, I was visiting a friend who happened to be a talented massage therapist. She was widely recognized for having a very strong, intuitive talent—a talent that I had witnessed many times while watching her help others overcome difficult emotional issues. We were visiting over tea, as we often did, but at some point in our conversation she stopped and said, "Tim, this may seem weird, but I have a strong sense about this. Would you please go over to that corner and get down on your hands and knees with your head in the corner." Perplexed, but trusting her and her unique way of experiencing the world, I carefully followed these directions. She moved up behind me and with one hand on my hip and another on my shoulder, she pushed hard, pinning me into the corner. Suddenly, a completely unexpected and torrential flood of fearful emotional memories and physical sensations surged through my body. I could not believe what was happening. It was as if a spillway to some dammed up part of me, a part about which I was completely unaware, had suddenly burst wide open. I recognized these surprising, but painful, physical and emotional memories as my own, but it was as if I was feeling them for the first time.

That event was the dramatic beginning to a long process of recovery, as I gradually opened to these hidden and very traumatic memories. Over time, they have slowly become more visual, verbal, and identifiable; I was finally able to address and release a great amount of fearful tension that I had been habitually holding. I know now I was a victim of childhood abuse at the hands of a neighbor, one whom I had deeply trusted. Looking back through my life, there were many obvious clues and hints that I and my parents had completely missed. Until this breakthrough, I had a complete loss of any memory about the multiple boating trips with this person, along with a nearly total loss of my memory about anything else that happened during that period of my life, between the ages of nine and twelve.

Other than the critical healing begun with the surfacing of this buried memory, I also learned, at a deep and personal level, just how effective our minds are at controlling our perception and memory of events. We remember and retain only what we are capable of processing— conceptually, physically, intellectually, and emotionally. If something is outside of our limits, we automatically engage what psychologists call *cognitive dissonance,* repressing the event and filing it in our body's equivalent of a "dead-letter drawer." This is one common method we utilize to quell any inner dissonance.

This event highlights how our conscious minds and senses filter or miss *information.* It also demonstrates how this filtered *information* is still stored and available, even if we normally lack direct access to the actual memories. This story also illustrates one more example of a person possessing a type of talent that Western science cannot technically explain. Since our Western culture typically distrusts these unusual skills and insights, we lack the cultural examples—

archetypes—that would help us process experiences falling outside of our consensual norm.

FIRE WALKING

In the mid-1990s, a friend invited me to a fire-walking workshop that she was attending. Having never participated in fire walking, I was extremely curious and gladly agreed to go. In the 1970s, while living in the Fiji Islands as a Peace Corps volunteer, I had repeatedly witnessed the traditional Fijian fire walking performed within rituals. I knew that the Fijians have very tough, calloused feet, and I always assumed that those calluses played a major role in their ability to walk on red-hot rocks.

Ten first-time participants signed up for this fire-walking workshop but, unlike the Fijians, all of us were tenderfoots. We spent the entire day in psychological preparation, going through visualization exercises and meditations. The preparation focused on assuring us that *"we could do this"* and on the importance of not having any self-doubt. We were warned that if we had any doubt or "second thoughts" at the threshold, it was not only acceptable, it was absolutely expected that we would "opt out" and not walk. There was no peer pressure to walk, since we all understood through our coaching that walking with any doubt would lead to burns, and this could be catastrophic for the entire group. As it turned out, several people in the group chose not to walk at the last second.

In preparation for the walk that night, an enormous pile of wood was reduced to a field of golf-ball-sized, red-hot coals that were then raked into the 30-foot long, four-foot wide bed. We were to slowly walk, not run, the full 30-foot length, barefoot. As we stood near the threshold and felt the intense heat, every one of us had immediate second thoughts; the radiant heat from the coals was unbelievably hot, even when we were standing back several feet.

The workshop leader walked first with no visible ill effects, and that inspired my friend to walk next, also without problems. Encouraged, I walked to the edge, cleared my head of all thought as rehearsed, and started walking. Time and sound stopped and I felt a dramatic, yet still, calmness as I walked the ten or so steady steps required to complete this journey. I arrived at the other side completely unharmed and ecstatic. I had not even experienced the peripheral heat of the red-hot coals on the rest of my body. Several others walked; several chose not to. Somewhere in the process I realized that I wanted a picture of myself walking, so I decided to walk again. This time, three-quarters of the way across the hot coals, an "outside" thought entered my mind; I suddenly realized that the photo had not yet been snapped. While looking for my friend with the camera, I lost my intense focus and, in that instant, my feet were suddenly and sharply "stung," as I felt the intense burn of the hot coals along both feet! Stimulated by the pain, I immediately was able to regain my concentration and complete the walk. After the walk, I quickly checked and found that I had a few small blister burns on the outer edges of my feet, but nothing that was proportional to the burning sensation that I felt or what should have happened in the "normal" world. Another walker also experienced small burns, and she confessed to lapses of concentration during her walk. For me, this entire process was a powerful demonstration of the power of focus and belief. I could not have planned this experiment in a better way.

AT ONE WITH THE WIND

I was a fanatic windsurfer in the 1980s, and because I preferred high-speed sailing, I would only take time off work if the wind was blowing over twenty knots. This meant that the wind was blowing fiercely whenever I drove out to the lake where I windsurfed—a forty-five minute drive from my house. My old Volkswagen camper van had a worn steering gear and, because in those days my extra cash always went into new sailing equipment, the loose-gear problem continued without repair. The wind would invariably push and buffet the flat-sided van, forcing me to weave all over the road; it was almost impossible to keep that van on the road during the trip out to the lake. The drive out was always very stressful, but fortunately in those days, that particular road had very little traffic.

Once sailing on the windy lake, there was always a point at which something would change deep within and I would merge with the waves and wind. Somehow, I would release conscious control to become an integrated part of those elements, free of thought and perfectly fluid in motion. This extraordinary ecstatic experience of oneness with the wind and water was the primary reason why I loved to sail.

The return ride home in the van would always be a different type of journey than the morning trip. Even though the winds were still buffeting my flat-sided van, I could always easily steer right down the middle of the road without any thought or struggle. Some part of myself had learned to predict and read the wind in a manner normally hidden from my conscious brain. I was able to anticipate and adjust for the blasts of wind in intuitive and instinctual ways that made no sense to my time-ordered, rational mind. During those special times, I was tuned to the movement and flow of the natural world through a type of connection that was normally invisible and unavailable.

MY FIRST ACUPUNCTURE TREATMENT

Throughout an eight-year period as a young adult, I had a persistent and chronic health issue that dramatically affected my life. During that period of my life, I spent a large amount of time and money on hospitalization and Western medical technologies. Facing life-changing surgery, I was blessed to stumble across a young doctor who had acquired some atypical experience with Eastern medical systems. He explained that Western medicine did not have good options or a history of positive results for my particular chronic condition. However, while serving as a military doctor in Vietnam, he had personally witnessed many cases where my exact issue was cured through acupuncture. Suggesting that I try acupuncture before resorting to surgery, he was careful to request that I not tell anyone about his recommendation! This was 1974 and the medical establishment did not yet accept alternative medical techniques such as acupuncture. He feared that he could lose his license for even suggesting this option.

Living in the South Pacific at the time, I easily found an old Chinese practitioner. As I described my problem to the acupuncturist, he laughed and said, "Oh…this is very easy. One treatment, maybe two at the most." In that precise moment, after years of frustration, I felt a strong palpable change, which, to this day, I believe was the actual moment of "healing." Once inside his office, I had a seven-point (needle) treatment, and two days later I felt perfectly well. My difficult and long-term physical problem disappeared and has never returned.

The confidence that the acupuncturist displayed totally changed my internal conceptual idea about how difficult my problem was to cure. Over the previous eight years, I had been witnessing a great amount of confusion about my condition from the Western medical establishment. This may have also included many "unheard" comments between medical personnel during three procedures, while I was under anesthetic and theoretically "unconscious." These memories were all probably stored deep within my subconscious and may have contributed to my emotional and physical stress. Years of this type of negative input had convinced me that my issue was difficult and incurable.

I clearly experienced a significant change during the conversation that unfolded before the actual acupuncture treatment. I am sure that the healing began with this alteration of my thought patterns about my condition, which was probably the acupuncturist's intention. The actual treatment had an additional beneficial effect because it was designed to increase blood flow to my inflamed organs so that my own body could, and would, take care of itself. However, it is clear to me that something significant changed even before the first needle. There was a dramatic shift in my attitude, brought on by the old Chinese acupuncturist's confident laugh when he said, *"This is easy!"* Those words altered my entire belief system about my disease and, in that moment, I became more open and able to shift "universes." I entered an alternate parallel universe where my illness was no longer lifelong and debilitating.

Today, there are numerous signs of change in the Western medical establishment's attitude about alternative techniques such as acupuncture. Hospitals now have acupuncture clinics; nurses practice massage and midwifery; and physical therapists now use a number of once-strange techniques such as Emotional Freedom Technique (EFT). Western doctors still cannot explain why acupuncture and some of these other practices work, yet they are now willing to prescribe some of these treatments. The awareness of the importance of attitude is also slowly growing. As a culture, we are apparently becoming more comfortable with the Eastern idea of "not knowing." This occurs as we also become more aware of ourselves as *quantum beings*, brimming with *infinite* possibilities.

SHARING AILMENTS WITH A LOVED ONE

My wife and I have always felt a strong connection, but after about five years of living together, we began to observe a strange and new phenomenon. At first, we noticed that we had the same aches, pains, and ailments—there was some type of intimate interconnection forming between our bodies. One of us would mention a physical issue only to find the other was experiencing the exact same symptom. At first it would come up only occasionally, but within a few years it was occurring with such regularity that we would simply say, "How's your left shoulder?" and receive an understanding nod. This once-strange pattern had become our norm.

With common conditions like muscular aches, the physical location of the ailment or pain is typically mirrored—meaning that the left side on one of us corresponds to the right side on the other. If the ailment is specific to a body part that is not bilaterally symmetrical, such as the gall bladder, heart or descending colon, then the actual parts and sides correspond.

This connectedness has helped each of us to become more aware of the other and now, when we massage each other, it is much more effective and interactive. We have also observed that when one of us clears an issue, the other also benefits. Again, I am struck by another example of the invisible connections that extend beyond these bodies: types of interconnections and interactions for which our culture and science currently have no clear explanations.

VOICE OF WARNING

In the late 1970s, I was a schoolteacher on the remote and rugged Kandavu island of Fiji. My students had a soccer match at a distant village, and since there were no roads or cars on this island, we had to travel many hours by boat. The school sponsored our journey because they considered this match very important. Since there were many people in this remote village who had never seen an outsider, I felt an added responsibility, knowing that I would be honored as a special guest. I represented my school, the Peace Corps, Americans, and all foreign visitors to Fiji, so I was especially careful to follow my well-rehearsed protocol.

During the match, I observed that the lemonade was made directly from river water and unfortunately, as I expected, the very first glass was offered to me. All village eyes focused upon me and anxiously awaited my approval. I held the glass up but immediately received a clear intuitive message, a powerful feeling or "knowing," that I was not to drink that lemonade; it was absolutely clear to me that if I did drink that glass of lemonade, the physical results would be significant. Throughout my years in Fiji, I consumed hundreds of drinks prepared from questionable water, but this was the only time that I was warned in this way. Aware of local customs, I knew that refusing the drink would be a major insult and create long-term problems for my school. I chose to ignore my strong internal warning and gulped down the lemonade. Ten days later, I fell into the early stages of the deepest, longest, and most painful sickness of my life; I had contracted typhoid fever.

The source of that "knowing" cannot be explained from within our traditional paradigm. I have learned to respect and trust this kind of deep intuitive knowing and, since that fateful day, I have been much more careful to honor these unexpected messages.

FIJIAN MAGIC FEET

When I lived on Kandavu, it was common to make nighttime treks to the neighboring villages for social events. Since this remote island had no roads or automobiles, boats and hiking were the only travel options. These hikes usually involved walking for a couple of hours up and down the rocky terrain of volcanic mountains, through rivers and marshes, over beaches and mud flats, and through almost every other type of possible terrain. Initially, because of my tender feet, I attempted to negotiate these trips in shoes or sandals, but I was constantly losing my footgear in the mud, tearing my sandals up on the rocks, and dealing with sand or pebbles. My personal shoe problems became the object of constant, but good-natured, teasing and, because I was slowing down the entire group, I decided to try walking without my footgear. After a few painful months of building the needed callous, I discovered that it had become very easy to traverse the complicated landscape barefoot, even on the darkest of the new moon nights. I noticed that I no longer stubbed my toes on the rocks or lost my footing by stepping on an unstable boulder; my feet naturally found their own solid footing. I developed the instinct and ability to adjust my weight instantly when my foot landed on sharp objects, as my toes spread wider and I could control each one more easily and independently, much like my fingers.

On one particular dark night, I started "thinking" about how amazing my feet's new ability was and immediately stubbed my toe. In that moment I suddenly understood. My feet had developed a radar-like instinctual sense of their own; they "saw" the terrain and "knew" just what they had to do without checking in with my brain. As long as my conscious brain did not get in their way and micromanage, my feet did a great job on their own. I really miss my sensitive "Fijian Feet." Not since those days of long barefoot walks have I felt so comfortable and connected to our Earth.

Our bodies have an amazing depth of intelligence of their own that is not yet widely understood. African drummers' hands have been measured to move much faster than our nervous system theoretically allows—the number of notes they can play each second is not supposed to be physically possible. New research indicates that the vast majority of our brain's processing occurs in the subconscious; what we are capable of understanding with our conscious brains is just the tip of the iceberg.

THINGS DISAPPEARING, THEN REAPPEARING

During several periods of my life I have observed the mysterious disappearance of small but important objects, only to have them strangely reappear, days later, in exactly the same spot where I "lost" them. I remember clearly the first time that I had a conscious awareness of this phenomenon. I placed my wallet on top of my desk as I walked into my office that morning, just as I did every day. At the end of the day when I went to retrieve it, the wallet was not to be found. Two days of frantic searching ensued before I accepted that it was inexplicably lost, having reached a point of a complete physical, emotional, and psychological release. When I arrived at my office the very next morning, I found the wallet sitting on my desk, precisely where I had left it three days earlier!

At that time, I did share an office suite with another architect, but since he was extremely serious about our practice and not prone to playing games or tricks that might waste our valuable time, I knew he was not the culprit. In fact, he spent much of his own time searching with me and there had been no other office visitors during that period. I assumed that there was some unknown yet "normal" way for this to occur, even though I could not think of one.

Then, this type of thing started to happen more frequently. The first few times it happened, I again ripped my office apart looking for my lost wallet or keys. Eventually, I began to shrug and laugh when it happened, knowing that my prized possession would return when it was ready. Over time, I noticed that my things came back to their spot only after I fully relaxed any *attachment* to these objects; I had to reach the point of no longer caring.

I jokingly began calling this recurring experience *"my poltergeist"* and ceased to worry about it. I recognized that it provided powerful lessons about the value, location, and impermanence of "things." That first wave of these "disappearances" lasted about six months. Ten years later, I experienced another series of "disappearances," and five years after that there was another brief recurrence. I quickly adjusted to these events and even started to enjoy them, as I began to see them as reminders or "continuing education." There was no simple, rational way to explain these occurrences, and I feel fortunate to have had these opportunities to observe actual physical objects directly expressing the same type of strange behavior that quantum particles exhibit.

GREEN SUIT

Even though I have learned to recognize and trust my "inner voice," I am still sometimes impatient and sometimes I judge the voice as insignificant or even wrong. On the rare occasion that I do ignore this inner voice, I inevitably receive a quick lesson about trust. This was the case on a particular day a few years ago, when I pulled up to my parking place at the large, public pool where I swim my daily laps. As I reached for my usual bright green swimsuit in the back seat, an inner voice clearly said "NO! Not the green one." This seemed completely crazy to my rational mind. "What could possibly be wrong with the comfortable swimsuit that I wear every day?" I had another suit in my trunk, but I was late and I did not like the way the other suit fit so, again, I just reached for my green one. The same inner voice repeated itself! I resisted, judged the voice as *"ridiculous,"* and grabbed my old green suit.

When I changed in the locker room, my favorite swimsuit felt curiously uncomfortable, but being under a severe time constraint, I fought the persistent urge to get the other suit. Feeling unsettled, I quickly entered the very crowded pool area and swam my laps.

Upon my return to the locker room, a lifeguard I knew pulled me aside to tell me there had been a complaint filed against me; I had been accused of a very serious crime. The victim, a young child, had identified me from far away on the other end of the large crowded public pool by pointing and saying, "That man in the green swimsuit!" Apparently, it was my bright green swimsuit that connected me to the crime—the perpetrator wore a similar suit.

Fortunately, because I was a regular at that pool, all the staff knew me, my family, and my habits. Additionally, because the crime occurred in the children's area and a large number of witnesses and friends could collaborate that I did not leave the lap area of the pool, I was cleared without any further issues. This process certainly consumed much more of my time than it would have taken to simply change suits when I heard the warning.

The entire incident was highly educational and thought-provoking for a number of reasons. I was impressed with the deep power and wisdom of this strange warning that seemed crazy and unimportant to my rational mind. I was also surprised by the unexpected turn of events that unfolded because of such a simple choice. We never can be sure how things will unfold or how the most unexpected twists and turns can interact to play critical roles. This experience also gave me the opportunity to experience life from a new and unexpected perspective—that of the innocent and wrongfully accused. This was a perspective that, before that day, I had never actually experienced (for when I was accused as a young boy, I was usually quite guilty).

This incident again reminded me of the expanded awareness and intelligence of our "inner voice," even though its message might seem illogical from our rational perspective. This is a "voice" that originates from somewhere beyond the boundaries of our rational mind and can speak to and about things that are completely unseen.

FEELING MY FATHER'S PAIN

When my father passed, I was sitting in my living room two thousand miles away. He had been ill for some time and I had just returned, that very evening, from a long and fulfilling visit with him. The actual trip was initiated by a very clear intuitive "knowing" that it was the right time for the final visit—a very common but rarely discussed human experience. His dying that very evening was still unexpected because when I left him earlier that day, he seemed renewed, and it appeared that he would still be with us for some time.

I was awakened in the early morning hours by a strong and painful burning sensation in my leg. The pain was excruciating, but not coming from a place in my body that had been troubling me. I realized that it was located in the exact spot where my dad had been experiencing a great deal of his pain, and I understood that connection instantly. With that realization, I had a very strong and clear sensation deep inside: I knew that my dad was passing at that exact moment. Within the hour, I received the call confirming his transition.

This type of close interconnection is very common and many people have similar episodes. Later I will relate another story about a friend's very dramatic connection with her twin sister. What is odd is that these common intercommunications are not more openly recognized, studied, and integrated into our science and culture.

BURST OF ENERGY AT TIME OF DEATH

My mother made a conscious decision to exit this life soon after she broke her hip during a fall at her own sister's funeral. Her last living close friend had just passed, and she could no longer participate in her favorite activities. Alert long enough to allow everyone in her family the opportunity to spend individual quality time with her, she eventually lapsed into a coma. Five days later, only after all her children assembled, she suddenly became extremely conscious and alert. She was unexpectedly with us again for the next three hours as she entered her transition and the moment she left us was absolutely clear.

This burst of energy at the end of a person's life is a common and well-documented phenomenon. Where does this shift in awareness and extra energy come from and why? It seems as if the spirit wants this time for closure, or to gather extra energy to complete the physical separation from the body.

Many parts of my mom's dying process reminded me that spirit can be independent of the body. In her last moments, my mom's spirit was present, full of energy, and shining, even though her body was completely worn out; they were clearly not dependent on each other in that moment. My mother taught me many things and this wonderful demonstration was her last great gift.

THE BIRTH OF MY CHILDREN

At the other end of life's bookends, I have been intimately involved with the moment of birth. Both of my children came with unique spirits that could be instantly recognized from their very first moments. At birth, their two spirits were as clear and different from each other as they are today—these same spirits continue to manifest and shape both of their individual life trajectories. With these first meetings, it was clear that my first child, Emily, was extremely wise and contemplative, but seemed to be already burdened with a heavy awareness of the human condition, suffering, and the complexity of humankind. From the first moment, my other child, Rachel, was wide-open for any experience, fully engaged physically, and always actively looking for something new to play with or explore. Over the years, their minds and bodies have evolved to allow for different kinds of personality expressions by their unique and different spirits, but the fully recognizable core of each individual has remained unchanged. When my children came into this world, each arrived with her own identifiable and expressive spirit already present.

Many, possibly most, parents have this same clear awareness; each child's most fundamental being is unique and present from birth. This presence will evolve, but it usually does not change in character as the child matures. Deeply integrated into their bodies and lives, it continuously shapes their growth. My experience as a father has fully reinforced the idea that each of us arrives into this physical existence with our unique core spirit already in place.

ARCHITECTURAL IDEAS

An important part of my job as an architect is to find a single solution that solves a project's many different problems. Solutions to this type of problem are always *dynamic*, meaning that if one *variable* or unknown is changed, everything else must also adjust and change to compensate.

To solve these complex architectural problems, the first step is to list and study the various issues and *parameters*. The amount of time that I spend on this stage varies with project size, and often for larger or more complex projects, it can last several days or weeks. After this initial period of learning, I quickly try different rational possibilities to see which ideas work and which ones do not. At this point, I can usually find logical solutions that work for most, but not all, of the issues, and any modifications that I try usually generate brand-new problems of their own.

Many designers reach a point in this process where it becomes clear that the "rational and logical" brain is no longer the best tool for processing all the complex *variables*. A computer, if carefully programmed, might solve some aspects better and faster, but the thought process of the programmer will always define and limit the possible results. At this early stage most of my designs seem rather flat, uninspired, or even boring—this can be the most frustrating step of the entire process because the perfect or ideal solution often seems to be impossibly out of reach.

Through many years of practice, I have learned that once I have thoroughly worked the problem, but just before I get frustrated, lies a special moment where I can consciously stop the rational design process and literally "sleep on it." What usually happens is that I wake up at about 3:00 a.m., often during that very first night. I wake with only a slight hint of an idea which, I have learned, I must immediately put to paper. I start sketching in an observant but still dreamy state, as the vague idea gradually becomes more solid. At some point, I sense a complete solution, even without taking the time to check all the possible variables. I usually check a few of the more troubling or resistant issues, and, once I have confirmed that these are indeed solved, I can finally go back to sleep, knowing that I have the beginnings of a complete solution. In the morning, I continue the process by transferring the ideas to my computer, where I can thoroughly check all the issues and relationships.

From experience, I have learned to recognize that special "feeling" that always accompanies the arrival of these successful solutions—a "knowing" that arrives well before the actual confirmation of the solution and always turns out to be accurate. Another interesting thing about this process is that while the final solution solves the complex series of interrelated problems that were specifically addressed, it often provides a number of unexpected "bonuses" or solutions to problems that were not even considered in the initial analysis. It is as if the solution comes from somewhere outside of my normal conscious and thought-bound mind to include aspects that I did not initially understand or consider.

I have learned to trust this process and it has always been consistent and very reliable, but I am unable to explain it in normal, rational or linear terms. I am not the only person who relies on such methods, for this type of experience is very common with other designers and artists of all types—some use the word *flow* to describe the feeling. This mysterious and magical process, which I have come to love and trust, is the aspect of design that continues to interest me and allows my work to stay fresh and exciting after almost thirt-five years of practice.

While many of my designs require solving a specific set of detailed problems, sometimes the real "problem" of a particular design is that there are too many choices or potential solutions that seem to work equally well. How, then, does one decide which path to take? I have developed what might be considered an unusual approach. I simply ask the land or building what it wants to become. While this technique is outside of any logical or mainstream-approved design method, it does work, sometimes amazingly well. Many artists, musicians, and writers rely on this or similar techniques. Until recently, "asking the piece of art" was spoken of only in the most esoteric circles, but today I observe much more openness about this unusual process. Just yesterday, at a design meeting, the interior designer said, "When I remodel a home, I always ask the house where the front door should be placed." About fifteen years ago, I added a slightly modified version of this technique to my design process. I began to visualize the already completed building at some date in the future. For me, this "mental time machine" has been a very successful method for finding creative and concrete solutions. Our rational and scientific minds have a very difficult time explaining how and why these strange methods work.

Gertrude Stein was a contemporary of the earliest quantum physicists. She was blessed with a mind that thought in ways that seemed strange when viewed from a logical framework, but not as strange when seen from the emerging *quantum* perspective. She was a perfect translator for her times, and she helped many artists and writers become comfortable with the process of being "open" and available to answers arriving from unexpected places. After one of her lectures, one member of the audience remarked, "Gertrude said things tonight that will take **her** years to understand." New and unexpected ideas seemed to pop into her head even as she guided others through this intuitive process.

During my training at architecture school, I was taught some of these tricks. However, I also was cautioned to explain my process to clients in very logical and rational terms because, in the words of one professor, "letting the clients see the inner workings of the creative process is like witnessing the making of sausage—it tends to erode confidence."

A technique called the *charette*, which involves a multi-day design session allowing no time for sleep, is a standard practice for inducing creative ideas in many architecture studios. The entire purpose of this exercise hinges upon exhausting or breaking down rational thought processes, along with *egos,* to better allow ideas to simply emerge or *flow* unfiltered from that "other" place. This process is physically and emotionally taxing, but very effective. The excellent results make no sense from our standard, three-dimensional, classical *viewport*—something else is clearly going on!

EXPERIENCES OF FRIENDS AND FAMILY

CLOSE CONNECTION BETWEEN TWINS

Forty years ago I was living in Wisconsin. My girlfriend's best friend was a twin and her twin sister, whom I had never met, was living in London at the time. One night my girlfriend received a call from her friend, who was in terrible pain and needed a ride to the hospital. All night long, hospital staff conducted tests, searching for the cause of her excruciating pain; but all the tests proved negative and the doctors were thoroughly baffled. In the morning, the pain suddenly stopped and she was sent home, only to receive a call within hours informing her that her twin sister had been in a terrible auto accident in London, survived for several hours, but ultimately died. The timing of the event precisely correlated to her pain the night before.

This type of special intimate connection between twins is well documented. In our culture, even though we recognize these unusually close relationships, we offer no clear explanation for the existence of these strong connections. Through the structure of the *Multiverse,* we are all intimately connected at deeper levels, but the connection between some twins seems to lie much closer to the surface. This incident, the experience I had when my father passed, and many others like these indicate the existence of strong invisible connections that cannot be explained from our old *viewport.*

DORM-MATE HAS AN ACCIDENT

In 1970, when my sister was a freshman in college, one of her friends was in a motorbike accident which resulted in a traumatic head injury. After a long recovery, her friend returned to the dorm, but she now saw people differently—everyone was now surrounded by a large halo of auric light. Each person's individual aura had different colors, and she learned how to sense from the shape and color what the person was like, and whether their intentions were selfish or loving. After several months, she also started to add small but accurate premonitions to her extrasensory talents. One night she awoke screaming, waking everyone else in the dorm wing. She had just had a vivid dream that her parents had died in a car crash on a mountainous road.

She called her parents and they calmed her by saying, "Honey, it was just a dream!" For three successive nights, she had the same dream and her piercing screams woke everyone in the dorm. Each time she called her parents until, finally, they promised not to take a vacation, which they had already planned.

A short time later, she received a call informing her that her parents had just died in a car crash in the mountains—they had intentionally misled her about their plans. She then entered a deep catatonic state and was admitted into a mental health institution. At this point my sister lost touch with this dorm-mate, and does not know what happened afterwards.

This type of *seeing,* while rare, is certainly not unique. A friend of mine, Michael Tamura, who is an amazing and very successful psychic, has always seen patterns of geometry and color. Through the years, he has learned to interpret the meaning of these patterns into words and actual physical events, both past and future. Once again, this type of ability cannot be explained through our current paradigm, yet it makes perfect sense in a timeless *Multiverse.*

VISITATIONS FROM A DECEASED HUSBAND

My sister's husband, George, was a larger-than-life character, and his untimely death caught us all off guard. He had an accident while playing with his daughter's toy scooter on a hill in front of their house. Immediately after the accident he appeared to require only a quick emergency-room visit, but a couple of days later he unexpectedly died in his sleep from a related, but undiagnosed, injury.

It is noteworthy that the evening before his death, some part of him "knew" because, even though he appeared and felt fine, he led the family in prayer focused on his having made peace with all things in his life. This was unexpected and made my sister very uneasy because he had never done anything like it before. We often seem to understand things that we should not be able to know from logic alone.

In the months that followed his death, my sister and her teenage daughter experienced multiple "visitations" from George. Many of these happened while the two of them were together, so they had instant confirmation from each other that they were not just imagining these strange events. At that time, my sister and her daughter had no belief or interest in the supernatural, and they never shared these stories with anyone because they were convinced that no one would believe them. It took direct questions from me to provoke them to relate these stories, and even then, they were reticent.

George had been a lifetime cigar smoker. After his death, they twice came home to fresh cigar smoke in the house. Once, they found a lit cigar in his ashtray on the table of his smoking room, and another time they found a cigar burning in his desk drawer. Theirs was a large, well-secured, private suburban home so no one else would have, or could have, been inside their home.

Another time, while my sister and her daughter were sitting in one room, his favorite Lou Rawls album started playing on his turntable in another room. Neither of them knew how to use his very complicated turntable system. Again, they had no houseguests and there was no other person who had access to their home.

Another visitation incident involved one of their unresolved marital issues. My sister had often tried to control her husband's gregarious behavior by hiding his glasses so he had to stay home. Her method thoroughly annoyed him and this was still a very active issue when he died. One day many months after his passing, she suddenly could not find her glasses anywhere in her home. She checked the entire house, even though she clearly remembered exactly where she had left them. A few days after giving up the search, she went to her office, which she and her husband had shared for many years. Even though her last trip to this office had been several days before losing her glasses, lying there, right in the middle of her desk, were her missing glasses and their case.

These unusual events stopped after a couple of years, but my sister still gets chills every time she thinks back to this period of her life. Again, there are no plausible explanations from within our existing paradigm but the *Multiverse* provides multiple pathways for these types of experiences.

HEALING IN GHANA

A dear friend went to visit her daughter in Ghana a few years ago. Her daughter was married to a man from Ghana and they both lived there in his village. During a festive celebration, my friend's daughter accidentally sliced the bottom of her foot. Since the wound was large, wide-open, and bleeding, my friend's attention reflexively went to motherly concerns such as infection and knowing that her daughter had not had a recent tetanus shot. Her concerns were certainly called for; medical care and hospitals were quite far away. The daughter, who had been living in this village, was adapted to the local ways so when the village "healer" noticed the injury and volunteered to "fix" it, she quickly agreed. After first holding her foot for a little while, the healer grabbed a handful of dirt and rubbed it directly on the wound. My friend's concerns quickly turned to alarm at seeing this, especially since she had observed the same dirt soiled by pigs and other livestock. The healer then pulled out a white powder that he always carried and sprinkled it on this now dirty spot, held the foot for a while, and pronounced it "all fixed!" My friend and her daughter immediately went home and washed the paste off, only to discover, to their amazement, that the cut had completely disappeared. It was as if this deep cut had never been there at all.

We have no mechanism in our *viewport* to explain this type of occurrence, yet my friend and her daughter had together just witnessed this "impossible" transformation.

SAVED BY A MIRACLE

Years ago, one of my very best friends was riding in the back of a camper pickup truck that was heading down a hill in New Mexico at a rapid rate of speed. As the driver lost control, hit a guardrail and started to swerve, my friend knew that he was in deep trouble and quickly reached a peaceful resolution, convinced that this was probably his moment to die. He said an abbreviated prayer and then relaxed into a protective fetal position. The truck flipped over and the next thing he knew, he was sliding down the hill, sitting on a small broken piece of the camper shell that was acting like a sled. He slid all the way down the hill and right into the soft desert sand at the bottom; his only bodily injury was a broken finger. Unfortunately, the truck's driver did not share his good fortune. Most people, with some age and adventure behind them, have several of these stories about close calls that must be more than just good luck. As we become more observant, we might realize that such miracles have always been a "normal" part of our lives.

DEEP INTIMACY OF MASSAGE

A friend of mine is a very gifted body worker and physical therapist who has helped me through several difficult injuries. We sometimes share stories about our more unusual experiences. One day, when he was noticing a particularly deep connection with his massage client, he casually looked down at his own hand as it was working on his client's head. He was shocked to see that from his perspective, his hand was buried halfway inside the client's head. Stunned and not quite knowing what to do, he paused for what seemed like a long time and then slowly pulled his hand out. Too disturbed to say anything, he resumed the massage. At the end of the massage, the client commented that it was amazing how she could not distinguish where her body stopped and his began. My friend just held his tongue, needing time to process his own observations. Dissolution of our normal physical boundaries is certainly not expected, but it is also not unheard of or impossible. As mentioned in the discussion about the size and density of the atom, our bodies are entirely built of just space and *energy*. From an awareness that understands this perspective, it might even seem surprising that this type of "soft boundary" interaction is not encountered and noticed more often.

PASSING THROUGH SPACE BETWEEN ATOMS

While sharing some of these stories with several friends, they recalled many similar experiences, but one story stood dramatically alone. A friend had been a passenger in a car driving down an interstate when, all of a sudden, a large tanker truck, coming from the other direction, jumped the divider and slid sideways right in front of them. My friend's automobile plunged directly into the tanker truck broadside at 75 miles per hour. His last momentary thought before the crash was a clear knowing that he was about to die, for there was nothing but a massive wall of steel directly in front of him. The next thing he remembered was standing completely unscathed on the other side of the tanker truck. The car was destroyed, pinned under the frame of the tanker truck and his car's driver was dead. He has absolutely no idea what occurred, how it happened, or why he was completely spared from any physical damage. The only explanation he has is that he somehow passed right through that truck.

UNIVERSAL OUT-OF-THE-BOX EXPERIENCES

SPORTS AND GAMES

While participating in sports or other similar activities, we can have internal experiences that reveal or predict the next series of future events. For me these types of occurrences are frequent, and through talking to others, I have learned that they are universal and quite common.

Precognition is an awareness of something that will be happening in the "future." For me, *precognition* occurs often in games and sports, especially in slower games like billiards, golf, and baseball where the longer pauses allow more time for thinking. While winding-up to throw a pitch, back-swinging the golf club, or lining-up the billiard's shot, a vision might pop into our minds and suddenly we "see" and "know" the entire shot before it unfolds. This "vision," ahead of real time, can often appear just like a movie clip of this perfectly executed pitch or shot, or sometimes, a movie of a very easy, yet completely blundered, shot; *precognition* seems to work both ways. As the action unfolds, the participant has this internal awareness or absolute knowing about exactly what will happen in the next few moments—the actual event simply confirms that previous mental vision.

The first few times that I noticed this occurrence, I was amazed, but through the years I became accustomed to these pre-visions. I eventually began to expect and encourage this type of vision so that I could experiment and consciously try to manipulate the outcome. Gradually I discovered something else; if I hesitated, second-guessed, or in any way wavered from the straight forward certainty around this vision, the shot could fall apart and result in a very different outcome—a second brief thought that questioned or analyzed the original vision had a disruptive effect on the outcome. Instead of simply receiving a glimpse of the near "future," I was trying to influence it and my involvement altered the outcome. Over-thinking and over-analysis can pull us out of the natural *flow* of events.

A technique based on this principle is now widely used in competitive sports. First developed during the 1960s for Olympic-level sports programs in the Soviet Union, it involves visualizing and meditating on the successful outcome as part of the preparation. By the early 1980s, friends of mine who were on the U.S. Olympic diving team were trained using these Soviet methods. They were instructed to imagine a movie in their head, one that illustrated a perfectly executed dive, and then they were to run this movie, repeatedly, in their minds as they prepared for their competitive dive. Today, this and similar techniques are commonly used worldwide in a variety of athletic training programs.

In sports, games, and activities that involve a significant amount of steady, rhythmic activity or quick reflexes, a different type of phenomenon occurs. A "knowing flow" or natural rhythm can sometimes take over and guide the process. When this occurs, it feels as if the mind has fallen completely out of the picture, as the body alone seems to act and react at an extremely high skill level. I have personally experienced this automatic response most often in tennis, music, and high-speed windsurfing. At the beginning of a session, I am usually thinking about exactly what I am going to do and why because, initially, my brain is fully engaged in the planning. Then, during particularly intense or challenging moments, the mental activity seems to fall away and, I begin to have the sensation of watching myself sailing, playing tennis, or performing a musical part, from a somewhat uninvolved, outside reference position. The body is still doing its job at a very high skill level, but after this shift, it seems to be able to operate without my conscious control; it is as if I am no longer directing my own body. If I consciously focus or think about what I am doing or how well I am performing, the mind's relatively slow judgments become involved once again, causing difficulties. This sensation is not unlike my observations during fire-walking or hiking barefoot. Much more goes on "below the surface" than we usually imagine.

UNDERSTANDING SUBTLE TYPES OF PRECOGNITION

Fully envisioning how a golf shot will unfold before the actual swing is one common example of *precognition* in sports. *Precognition* occurs throughout our lives. It might appear as a clear vision, but more often it materializes as a strong sense about something that is about to happen. Most of us have had that sudden and strong intuition that we should or should not do something, but we usually have no clear idea about the source of this *information*. Impulses like, "I should not get on that bus" or "I should walk down that block instead" are shared by all of us. (I have already cited several examples from my own life.) It can be impossible for the unexamined mind to determine whether this impulse is originating from a deeper level of timeless *information,* or whether it simply is generated by our conditioned, but unrecognized, fears.

What varies from person to person is how each of us trusts these "out of the blue" impulses, and, therefore, how we respond. Our contemporary culture teaches us, in multiple ways, to ignore these impulses, since they are not testable or explainable and, therefore, are not considered real or accurate. If over time we pay attention to and experiment with these intuitive impulses, our discernment and trust in them tends to increase and we become more sensitive to these and even more subtle messages.

There is always the problem of distinguishing the real warnings from our general subconscious fears. What we are receiving is only the shadow of an energetic impulse containing *information* that does not originate in our *viewport*; this *information* is not always easy to interpret. What might appear to be a "warning" could instead be originating from some hidden, fear-based darkness.

If we have been leading unexamined lives while continuing to embody our egoic fear-based shadows within, then clear discernment becomes impossible. On the other hand, if we have been dedicated to clearing the darkness within, then we will have learned how to better distinguish the deeper impulses from those that are just ego or fear based. ***Exploring and clearing buried memories and emotions is critical to being able to distinguish our hidden personal fears from valid intuitive information.*** Through this expanded form of wisdom, our intuition becomes much more accurate and then we, in turn, naturally begin to pay more attention to our deeper insights.

DÉJÀ VU

Most of us have experienced the feeling or sensation commonly known as *déjà vu*—the French expression for "already seen." *Déjà vu* often appears as an intuitive sense or "knowing" that we have been in the same place or situation before, saying the same words to the very same people. It can appear as the strange and sometimes eerie sensation of having already lived that particular scene in your life. Sometimes this common form of *precognition* also includes precise knowledge of exactly what will happen next. *Déjà vu* is another relatively common experience that cannot be explained using the "normal" rules governing our familiar space and "time."

However, within the larger and expanded framework of the *Multiverse,* there are many plausible explanations for how events of this type could unfold this way—the timeless architecture of the *Multiverse* can easily explain this mystery and similar phenomena. From a broader perspective, one where all time unfolds at once, we can understand *déjà vu* as a simple glance forward in time or, possibly, it is a brief connection to a similar parallel universe. *Déjà vu* is another of those common and universal experiences where the "arrow of time" and the bounds of space do not follow our usual three-dimensional expectations.

MANIFESTING A FRIEND

Another related occurrence is that of finding ourselves thinking about someone and then suddenly "out of the blue" we see them, hear from them, or hear something about them. We might be thinking of someone we have not seen in twenty years just as the phone rings with news about them; or wondering about an old acquaintance, we turn on the TV to see a news article involving this friend. While we can explain some of these moments through pure coincidence or simple probability, the frequency of and timing around other occurrences will completely defy normal probability. Also noteworthy is the fact that these occurrences often happen when something important is happening in the other person's life; it is as if we feel their worry or concern.

Again, this type of occurrence involves the breakdown of one-directional marching time or the penetration of the invisible membrane that separates us as individuals. This type of event speaks to the deeper connections between all of us.

KNOWING THAT SOMEONE IS STARING

Another very common experience is to be sitting somewhere and "feel" another's attention directly upon you. We have all had this sensation and nothing within our standard five senses explains it fully. Depending on the circumstances, this can be a welcomed or a very disturbing feeling. Again, this demonstrates a familiar type of energetic communication that is not yet well understood within our realm.

ALARM RINGS ON DREAM CUE

Most of us have had the memory of being lost in a long and complex dream that culminates in some event or sound that magically blends with real world circumstances just as we gain waking consciousness. The waking world alarm rings right on its dream cue; someone knocks on the door in both the dream and waking world; or the house shaking in the dream blends into an actual earthquake. When this happens, the real event is choreographed exactly to the dream; our waking completes the dream without a loss of time or beat.

This phenomenon can be explained many different ways using our expanded new understanding of the architecture, including, most simply, a manipulation of our sense of time passing so that the entire dream was created and played out in the first nanosecond of the waking world event. No matter what the explanation, this type of phenomenon points to the thin boundary between our dreams and waking life and the fact that time is relative, malleable, and always a function of our perception.

CHANNELING

Human history has numerous examples of an extraordinary and widely recognized phenomenon called channeling. Channeling usually is defined as "the esoteric process of receiving messages or *information* from nonphysical or extra-dimensional *beings* or spirits." Frederic Nietzsche was sometimes involved with this process, as were other notables such as Edgar Cayce and John of God, who is still alive and practicing in Brazil. Helen Schucman, a professor of Medical Psychology at Columbia University in the 1960s and early 1970s, is said to have received the entire 1,000-plus-paged text of the *Course in Miracles* through the somewhat involuntary, but direct, channeling of Jesus.

One of the most intriguing aspects of channeling is that the *information* sometimes is relayed in foreign languages or at a technical level that the person acting as the channel could not have understood. They sometimes have no technical or cultural understanding of the material channeled—it might come through in a language unknown to the channeler, contain technical or historical information that they never could have known, or the information might be just "over their head."

REVELATION

Revelation can be classified as a specific form or subset of channeling. The main difference is that most of those receiving revelation do not profess that another is speaking *through* them so much as speaking *to* them. Many of the world's major religions were initiated through revelations received by their founding saints or prophets.

The Church of Latter Day Saints, one of the fastest-growing religions in the world, originally was founded in the early 1800s upon claimed revelations delivered through golden tablets in upstate New York. The LDS Church continues to establish new splinter branches as various individuals within the church continue to have their own personal living revelations—a practice that is permitted, and arguably even encouraged, by church doctrine.

Almost every religion in the world has roots and history based on some form of revelation so, at some level, we accept, and even revere, this form of channeling—at least for certain historical figures. Again, while there is no clear explanation for this phenomenon that makes sense in our three-dimensional realm, channeling and revelation are very easy to understand and explain within a dimensionally expanded *Multiverse*.

NEAR DEATH EXPERIENCE

One particular phenomenon "outside of our conceptual box" that has surged in public awareness during recent years is the *NDE*, or *near death experience*. The *NDE* occurs most often when an individual is on the brink of death, usually as a result of a physical trauma such as an accident or surgery. Those having the *NDE* have a number of common sensations and visions that are associated with this process.

They may experience that their consciousness and visual perspective is suddenly located somewhere outside of their traumatized body. It is not uncommon for individuals having an *NDE* to witness themselves from some remote location. They might observe their body in a mangled car below or lying on the operating table at the hospital, or they might be watching their grieving relatives in the hospital waiting room and even find themselves sitting in their family living room. People having *NDEs* have been reported to recall entire conversations that occurred in other rooms or remote locations and they sometimes overheard these while their body was technically unconscious, under an anesthetic, or even *flat lined* and officially brain-dead.

NDE events seem to span age groups and cultures, but they often include specific flourishes that tend to be choreographed for the individual's religious or belief system. *NDE* individuals frequently sense their conscious awareness traveling down some form of tunnel, and they sometimes have visions filled with various types of light. They have a range of unusual sensations and experiences that can include meeting deceased loved ones who offer comforting words or advice.

At some point during their *NDE*, the person may realize that they are being offered a chance to return to their body to finish some work and continue their life. Obviously, since they lived to speak of their experience, all the individuals interviewed about their *NDEs* managed to return to their original bodies and lives. After returning, they often experience a broadened sense of peace and carry a new type of wisdom or knowing, even though sometimes their new life can also involve inhabiting a badly compromised body. NDEs speak of leaving fear behind, in its many forms, and almost all who have had this expansive encounter return with a renewed appreciation for life and a clear knowledge that life continues beyond our physical death.

Parts of the *NDE* may be nothing more than a hallucination or the way the mind's chemistry works when we are traumatized. However, some aspects clearly involve the edges of our paradigm because, as discussed, *NDEs* will recount events that occurred in and around their bodies and return with *information* that they should not have been able to hear or know. The recent increase in public awareness of this phenomenon may be largely because the number of people having this experience has dramatically risen as modern medicine has become very skilled at resuscitating those who otherwise would have died.

OUT-OF-BODY EXPERIENCE

My journey to Whiteface Mountain while meditating was a classic example of a spontaneous out-of-body experience. In that section of this book, I also related other out-of-body experiences of several close friends. Some religions and spiritual practices speak about the ability of consciousness to travel or sustain itself outside of the physical body. There are even specific spiritual traditions, one called "Astral Traveling," that teach a very conscious practice designed to induce this altered state of being.

Many of those who have had *near death experiences* describe very clear, conscious, and dramatic out-of-body perspectives as part of their larger experience. It becomes clear to anyone, like myself, who has gone through this type of event that the body is not necessary for consciousness, at least not in the short-term. Once again, we are reminded that we are not just these physical bodies.

CROP CIRCLES

Since 1970, more than 10,000 crop circles have appeared in farmers' fields around the world; the majority of those seen and recorded have appeared in England. Similar crop circles and other unexplained Earth sculptures, depicting strong and powerful geometric forms, have inexplicably appeared throughout much of our known history. While we have no historical record of their construction, many of these more ancient structures could have been created and built by man. Recent crop circles are quite a different story because we have many living witnesses to their sudden appearance, but no proven witnesses to their actual construction.

From my perspective, that of a skilled professional builder of large geometric forms in the natural landscape, this widespread phenomenon requires an explanation that, at least partly, lies outside of our normal rational logic and science. Given the constraints of time, place, and technique, I find it impossible to conclude that people could have built many of these recent crop circles. These unusual and very complex geometric forms are often extremely intricate and very difficult to construct, yet they typically appear very quickly, usually overnight. Sometimes they manifest in a short window of no more than a few hours, and occasionally even appear within a few minutes.

Skeptics often claim that crop circles are created by pranksters beating the grass down with either sticks or their feet. There certainly exist examples of crop circles, some very crude and some more sophisticated, that have been, or could have been, man-made. Since 1996, a group of builders called "the Circlemakers" has turned the construction of crop circles into a commercial enterprise by making medium-sized and relatively simple formations that sometimes include corporate logos. It is also quite likely that other, even more sophisticated groups have created and constructed a number of these crop circles.

However, having been trained and intimately involved with our most-advanced human construction techniques, I am more than convinced that many of these crop circles have characteristics and qualities that indicate an unexplained origin. In my professional opinion, the timing, precision, and technical issues often completely rule out the "hand of man." Ironically, the crop circles that are actually known to be constructed by man clearly demonstrate the characteristics and limitations of man-made circles; these examples only illuminate the impossibility of other more complex designs being man-made!

The first step of any construction is called *layout:* the marking of the critical points and boundaries of the new design on the raw landscape. Designs that have right angles and straight lines have clear corners that are relatively easy to measure, correct, locate and accurately mark. Even with the simplest of *layouts*, after checking and rechecking measurements, adjusting, correcting, driving new stakes to re-mark the corners, and then stretching string lines to guide the builders, the building site is always left in a thoroughly trampled condition. There is enough for the layout crew to worry about without the additional burden of having to think about where to place every footstep.

To accomplish the overnight construction of a complex crop circle, all of this coordinated activity needs to be completed very quickly and in the dark, when workers cannot easily see and communicate with each other or even see the terrain itself. The *layout* for many of these crop circle designs are several orders of magnitude more complex than any of my house designs; in addition, they are usually built in the night and without the help of lights. In some examples, extremely large formations have been known to appear in less than an hour's time between two scheduled airplane fly-overs.

Modern, expensive computer survey equipment, called *total stations*, makes the *layout* process much easier since all layout points can be shot directly from one spot using a computerized system. While these instrument systems have been developed, and improved upon over the last twenty years, many very complex crop circles appeared long before their availability. Even with the *total station,* at least one roaming person must physically move around the site to mark and stake every critical point. Large or curved formations would require a substantial crew of workers. Doing this quickly (especially in the dark to avoid discovery) and then leaving no trace of footprints is simply not possible. Some of these formations are as large as one-third of a mile wide and yet, to my knowledge, no crop circle crew has ever been accidentally discovered during what would be a complex and chaotic building process.

The crop circle sometimes called "Swirl" is enormous and complex. About one-quarter mile in diameter, it appeared on Milk Hill in Wiltshire in 2001. To construct it conventionally, in a single night, without lights, is simply impossible. When viewed from near the ground, the complexity and scale of the task of finding and staking the center of each circle within the field becomes very obvious. Notice that there are no human footpaths. For scale notice the person standing in the center of the second photograph. (Images courtesy of Steve and Karen Alexander)

When massive crop circle Julia (1,000 feet wide) materialized beneath the gaze of Stonehenge in 1996, a group called "The Circlemakers/Team Satan" claimed that they constructed this fractal-based pattern, despite the fact that two pilots, a security guard, and a gamekeeper all claim that the formation appeared within a fifteen-minute window on that Sunday afternoon. It subsequently took a team of eleven surveyors more than five hours just to measure the design, and they were not worrying about errant footprints. With another crop circle that appeared overnight, the surveying company who analyzed the site quoted that it would have taken a minimum of five days just to mark out the starting points before beginning any construction.

Those crop circles known to be hoaxes or corporate projects all have something in common: a relatively simple, underlying geometry. These simple geometries require far less layout, but typically, even with this advantage, they are filled with inaccuracies, mistakes, and errant footprints. Clever builders could carefully plan methods to lay out and construct some of these more complex forms without errant footprints, but it would take a well-coordinated and rehearsed team many days—or even weeks—to carefully and accurately follow the necessary instructions needed to complete some of these designs. Many of the most spectacular crop circles are documented to have been created in mere minutes without witnesses seeing any lights or human activity in the fields. Their rapid appearance, especially at night, seems well beyond our present-day technical capabilities, even with the most advanced digital survey and marking techniques.

222

Something very important is unfolding in full view of the entire world; I find it interesting that more scientific attention is not directed to this phenomenon. This may be an example of *cognitive dissonance* on a global scale.

SPONTANEOUS HEALING

We all know or are aware of people who were very sick, given a short time to live by doctors, and then miraculously returned to good health. Sometimes diet, lifestyle change, medicine or treatment can explain the change, but other times these factors alone cannot account for such an unexpected and dramatic change. Attitude and something else, which is not easily explained within our system of knowledge, must also be involved.

While unexplainable within the old paradigm, there are numerous pathways within the *Multiverse* paradigm that can easily account for this type of spontaneous healing. Since all possible outcomes already exist in the interconnected Web, sickness and health always exist together, simultaneously and equally in parallel universes. Radical healing may simply involve learning how to shift our consciousness to a healthier universe.

SYNCHRONICITIES AND SERENDIPITIES

Synchronicities are the occurrence of two or more events that are not causally related and therefore unlikely to occur together by chance. Carl Jung first formally defined synchronicity as "temporally coincident occurrences of a-causal events." They can also be described as powerful coincidences, which cannot be explained through any cause-and-effect pathway that we already understand, and which occur together in a very meaningful way.

A common example of a *synchronicity* might be when a question that has just arisen in our mind is immediately answered by another who was not asked the question, or by something else, such as a TV show, the book we pick up, or a sign along the road. *A synchronicity* might manifest as a series of babies in one family being born on the same significant date, or seeing your high school yearbook for the first time in 20 years only to receive an email from someone in your high school class in the next moment.

Serendipity is often defined as a "happy accident" and is a particular subcategory of *synchronicity*. *Serendipities* often express themselves as something wonderful or someone special showing up, right on cue, in a person's life. My collision with my friend in London and my discovering my other friend in the hold of the fishing boat on Vancouver Island could both be seen as *serendipities* and *synchronicities*.

Last year, a long-lost, Louisiana high-school classmate of my wife, Connie, ran into some friends of ours in California. Through conversation they discovered that they all knew Connie, so our friends gave this old classmate her email address. Living in east Texas, he wrote and asked if he could visit her later in the year when he was scheduled to be in Austin for a conference. They set a date.

Just one week before this long scheduled meeting he was in central Kentucky for another business function where he picked up their local magazine; the cover displayed a photo of Connie. The magazine had interviewed Connie earlier because she was the featured artist at that Kentucky town's popular art show, a show that was coincidentally scheduled for that very weekend. He spontaneously attended that art show and they reconnected, halfway across the country, only one week before their pre-planned reunion in Texas. All of this unfolded quickly, after not seeing each other for more than 40 years. This sudden crossing of paths from multiple directions is a clear example of *serendipity*.

These types of occurrences are familiar to all of us. Usually we write them off as a product of chance or luck, but there are times when something greater than just odds or good luck seems to be involved. For me, these occurrences are another indication of a deeper order and *interconnectedness* that lies below the visible surface of our lives.

WORKING WITH THE DYING

Hospice workers, EMTs, and others who regularly interact with those actively engaged in the dying process can often share a collection of personal experiences and stories that we cannot easily explain within the standard framework of our paradigm. While it is likely that some parts of these episodes may be the direct result of the dying mind playing biochemical tricks on our consciousness, a certain, wonderful consistency to these stories points towards a continuation of our existence beyond the physical. These stories often include insights into the dying process itself, and speak to an expanded vision of life and spirit. With one foot in another world, sometimes the dying can hold the door ajar for the rest of us.

My mother was a heavy smoker and, unknown to her family, had significant stress and worry about this habit near the end of her life. It was not that she wanted to quit—her worries were of a different nature. I found her unexpectedly relaxed one morning during her final days. She related that she had a nighttime visit from "others" who had to come her while she slept and said, "Don't worry. There are plenty of cigarettes where you are going next. You do not need to worry any longer about not being able to smoke." This may seem silly to many of us, but it was exactly what she needed to hear at that moment to help her relax into her own dying process. Sometimes those dying will inform hospice workers that they have been told, by "nice visitors" or long-dead, but close, relatives, that they have come to help escort them on their journey.

The dying process often exposes or dissolves some of the hard edges that contain our paradigm. In our culture, we seem to have a universal fear of death, often regarding it as the ultimate failure. Like so much else that we fear, we usually avoid addressing the subject of death directly and honestly; we often push any thoughts or discussion about the process far away from our daily lives. Because our culture has this deep, inherent fear, we miss many of the wonderful, potential gifts that this important milestone in every individual's life has to offer. The biggest shared gifts are the frequent and clear insights about our connection to something greater and the continuation of our lives beyond our present bodies.

CONCLUSION

We have all heard multiple stories very much like those above, but we can find them uncomfortable or unsettling, so in our culture we usually avoid discussing them publicly and openly. We lack the vocabulary or the conceptual space to understand or communicate their meaning, and our personal fears can complicate this type of exchange. It is often simpler to avoid thinking about these difficult-to-explain experiences so, when things do not fit easily within our known worldview, we find and partition special, hidden places to secretly file these memories away. Thereafter, instead of providing insights, these unexpected events can actually reinforce and deepen existing energetic blockages and inner darkness.

These types of experiences are always much easier to understand in the full context of the *Multiverse* and, once again, they can all be explained through the deep, interactive interconnectedness of the Web, the energetic nature of matter, or the dissolution of the arrow of time and the membrane that separates us as individuals. "The proof is in the pudding." Our own life experiences in this realm point to the existence of a much more expansive universe.

SUMMARY POINTS

All of the following are different ways to describe a single principle. As with the division of time and space, these ideas are divided only to allow for the way our minds work.

- *This world, exactly as it appears before us right now, is absolutely perfect. It is perfect because it always precisely reflects our current state of being. It is a flawless mirror. Nothing is ever wrong or broken.*

- *We inhabit a three-dimensional structured realm that we call our universe. This frame of reference purposefully limits the extent of our conscious connection with the comprehensive, interactive, and infinite continuum that is the full Multiverse. The Multiverse is much more expansive than we could ever imagine; the vast majority of it lies far beyond our senses, instruments, and rational comprehension. At this stage in our evolution, we do not possess the conceptual or physical tools even to begin understanding its true depths.*

- *We are generally unable to see, experience, or understand what lies beyond our three-dimensional conceptual space. We live in a kind of "dimensional fog" or veil. This is not a problem that needs to be corrected because it serves us well by allowing us to focus on our present purpose and work. We all have occasional occurrences in our lives when we partially penetrate this fog. These "out-of-the-box" experiences allow us fleeting glimpses of what lies beyond.*

- *The three-dimensional space that we occupy is only one sub-level or realm of our Multiverse, which is constructed from at least eleven dimensions. All dimensional levels (realms) are completely interrelated and functional parts of the whole, and there is no hierarchy or order of importance between different dimensions. Every part of the Multiverse is equally important and necessary for structural stability, including our familiar three-dimensional realm. Each and every dimension has a critical role in creation.*

- *At least ninety-nine percent of our physical, three-dimensional universe that lies within our physical "cosmic horizon" cannot be seen directly with our eyes or instruments. Even "empty" space has mass (and therefore energy); in fact, almost all of the mass of the universe that we can measure is found in this "empty" space. Scientists have no real idea what this is, or why it is there! This means that even our three-dimensional physical universe, the part of creation that we understand best, is mostly hidden and mysterious to us. Mystery lies everywhere we look.*

- *Everything that we see or experience in our three-dimensional world is just the projection or "shadow" cast from other dimensions. None of what we experience is solid or real. Thinking of our world as a dream is much closer to the truth.*

- *Everything in the physical universe can be reduced to a discussion about the movement of energy in some form. Matter is only energy expressed in a different form.*

- *Time, as we understand it, does not really exist. We only construct our concept of time to organize our thinking brains. We would be completely overwhelmed if everything unfolded as it really does: at once. Time slows the unfolding down for us. Time is our friend and helper.*

- *Everything that did happen, could have ever happened, is happening right now, or could happen in the future already exists within the infinite and intimately woven fabric of the Multiverse. Every possible choice or option is always present and available to us within a "cloud of possibilities." Anything can happen, absolutely anything! Our presence, focus, thoughts, attention, and our always-evolving core vibration causes this "cloud" to "solidify" around one of these realities. Only then does the potential outcome suddenly appear as "real." The deeper truth is that this expressed outcome is no more real than any other possibility.*

- *In our Multiverse each present moment informs the "past" and the "future" equally through waves of information that travel in all directions; therefore, we have the ability to alter both our "past" and "future" equally with our actions in the present moment. The implications of this are enormous. Our past history and experiences are only those that we have "chosen" to remember—we have the potential to "choose" differently.*

- *At the most fundamental level, the Multiverse is built upon vibration. The expression of form, as we understand it, is only secondary.*

- *In our deepest place, the source of our individual being, there resonates a beautiful vibrational "song." We can think of our lives as vibratory symphonies resonating through many dimensions. We all play our individual parts, each of us being one instrument within the grand orchestra that produces the unified symphony that is life.*

- *Our individual but culturally influenced vibratory "song" determines exactly what part of this infinite Multiverse we interact with in any moment. What we see and experience is entirely determined by how we vibrate within, which is always only a reflection of who we believe we are in each moment. As we change how we resonate by changing our deep core thoughts and beliefs, especially through illumination of the darkness within the subconscious, the outside world will be seen from a different perspective: it appears to completely change. <u>We can only change the world by first changing ourselves.</u>*

- *Thoughts by themselves are not enough to create this type of change. To change ourselves and, therefore, our world, there needs to be a deeper shift at the level of our resonant core vibration: our personal, inner "song of life." Although this process of deep change can begin with a single thought, we can never achieve the deepest types of change at the level of thinking, alone.*

- *Our life purpose is to gain experience so we may evolve to understand and eventually embrace "all that is," and thus discover the deeper truth about who we really are. In our realm, we learn at this deep level only through direct experience. All types of human experience—all the "bad," along with the "good"— are necessary for the completion of this purpose.*

- *We can always influence the appearance of our reality, for at some level we do actually "create it" by shifting our viewport with every thought, breath, and feeling! Every personal experience is a direct result of the expression of our core vibrational state. As we evolve and drop our resistance, the universe does not change, but our access to new parts of it and, therefore, our experience of it does.*

- *Each of us—everyone else and everything else—exist in an infinite number of "parallel worlds" simultaneously. Because of the way space, time, and consciousness work, we are usually only aware—or think we are only aware—of one of these existences. All these different "selves" are fully connected and continuously share information. This exchange of information between all "selves" and all things in the Multiverse appears to be instantaneous, but it actually occurs completely "outside of time."*

- *Parallel worlds and universes can be thought of as stacked, so that identical or similar worlds and universes are always directly adjacent and accessible—they actually blend with each other perfectly. All universes, even very different ones, are intimately interconnected. This quality, where all parts of multi-dimensional space are always directly adjacent to every other part, is called enfoldment.*

- *Our consciousness is able to travel between these parallel universes, and we continuously and constantly do so. While living, we are always dancing through slightly different but always-adjacent universes; along the way, we may notice that our experience of the "outside" world also shifts. This is because the universe that we are experiencing right now is actually not the same as the one we occupied earlier; some of the players and scenes will have changed. Most of these "other" universes that we visit are so similar that we usually do not notice the differences unless we pay very close attention!*

- *Information about the Multiverse and everything in it is stored holographically. This means that complete information describing everything is expressed everywhere and is instantly accessible from anywhere. This interconnectedness between everything in the Multiverse is absolute, yet it is impossible to explain from our three-dimensional perspective or viewport.*

- *All things are alive with the quality of beingness and interconnected through vibration. At some deep level, all beings and things are so fully connected that it becomes clear that we really are all different aspects of a single "Being." At another, even deeper level, everything is only "one"; there is no awareness of any separation. At the deepest level of existence, there is not even that "one," for ultimately there is "No-Thing," or that which is forever birthing the Multiverse and all that it contains. Some might prefer to name this "God."*

- *The full embrace of everything in existence, including the good, the bad, the happy, the sad, the saint, and the sinner, is a critical and life-changing (and world-changing) step along each of our personal paths. Pushing anything away, especially the unpleasant, creates a reaction that only makes its effects even greater. Instead, if we pull it in, accept it, forgive it, forgive others, and forgive ourselves, then we not only heal ourselves but we also help heal our world. The key to transformation is to integrate rather than segregate; we grow by expansion, not exclusion. This is our most direct pathway to freedom.*

- *In the Multiverse, there is no such thing as death; death is a time-based phenomenon. Time is not real in any absolute sense—it is only a human concept. If "time" is not real, then what becomes of that moment in "time" when we die? The only death ever experienced is the death of our old concept of time. Death is also form-based, and we now understand that form is also not primary, lasting, or real. Our form is only what our experience of a deeper truth looks like as it passes through and vibrates within the three-dimensional realm. Our familiar world of form is only the projection or shadow of a deeper Truth that is cast onto three-dimensions.*

- *Through our deep inner resonance, we always "find," or are attracted to, the perfect experiences for the growth of our being in each and every moment; this includes the moment of physical death. When we "die," we drop our physical bodies, along with all the physical blocks that have accumulated within our egoic bodies. Our soul then naturally and freely flows through the Web to emerge at its perfect resonant place within the Multiverse.*

- Since the appearance of our world is always an accurate reflection of our inner vibration, everything that we see and experience is always only a reflection of what and where we are within our own process of evolution. The appearance of the world is our reliable guide and reflection and never a problem that needs "fixing." Knowing, trusting and being comfortable with this is an important key to personal freedom.

- Every new experience adds to both our individual being and the greater collective Being. No matter how any experience looks or feels in the moment, it is always a positive experience for our growth and evolution. This even includes those events that our physical body does not survive.

- Our dimensional level is designed for the celebration of physical form. Certain types of growth and evolution can only occur through this physical form. The work we do in our bodies is critical to the greater evolution of Being. This three-dimensional realm is both the place where we contribute to creation and the perfect gristmill for our individual and collective soul.

- Once we, as individuals, become comfortable with how life works, much more dramatic personal shifts can occur. We then become more free and able to play within our Multiverse and our beingness. We can fully participate in this play while we are still in our bodies, and this type of play turns out to be very expansive and fantastic fun!

- The world that we see and believe is "out there" is created entirely through the filtering and interpretation of our three-dimensional mindset, along with our cultural and personal concepts. Collectively, all of this can be called our viewport. When we adjust our viewport, everything that we think of as "out there" will change. This is actually the only way to change the nature and appearance of the external world.

- Everything we are experiencing right now is from our personal perspective. It is always subjective because our minds separate, filter, process, and then remember it in ordered time. It is our story! There is no such thing as an objective story at the three-dimensional level of human expression—everything experienced is always subjective!

- *Therefore, there is no point in trying to change the behavior of "others." In the physical realm, these "others" will only change the nature and appearance of their world when they are fully prepared! Besides, there really are no "others." We only need to change ourselves to change how all these "others" appear in our world. The world, as we see it, is our dream.*

- *We can only see that which conforms to our worldview and viewport. We change "our world" only by expanding the scope of what we are prepared to experience. We do this by discovering, embracing, healing, and integrating all that we resist.*

- *From our three-dimensional viewport, our journey towards Truth is gradual and evolving. What is true for us today will not necessarily be true tomorrow. As our perspective or viewport changes, so does our truth. In our human realm, everything is impermanent—even our truths. If we hold on to relative truths beyond their usefulness, then they become barriers that prevent us from growing or evolving. Being able to shed old ideas and being open to new levels of truth is critical to the development of our ability to dance freely within the Web.*

- *Because of the geometry of the Multiverse, if we do what feels "right" in every moment, our lives will unfold beautifully and fluidly. Discerning true, or "right," feelings from our old, habitual, fearful, and self-destructive responses is the critical key to this way of being. Clearing the shadows and fearful blocks is where we all must start; only then can our own feelings become our most trusted advisors. Throughout this process of self-discovery, we learn to recognize that there really are no "others," and act accordingly.*

- *We each are living unique, but constantly evolving, lives within the Multiverse through birth, death, living, thinking, interacting, vibrating, and being. If we do not feel right about what we are experiencing, we can change it through a process that begins with examination of our core beliefs, fears, resistances, and deep inner thinking. This process must include the full embrace of whatever we are experiencing right now because our personal experience is always the direct and* resonant reflection of our vibration in each moment; we can never hide from our true selves. Through this work, our inner resonance will change and, subsequently, our experience of the universe will change.

- In the end, there is really nothing to do except to be consciously aware of each "present moment" and explore any resistance. Watch, observe, and be a passerby. If we become a "hollow being" while practicing openness, compassion, and love, then freedom and joy will follow naturally.

- Living our lives with an awareness that we are all interconnected makes life more fluid and joyful. It is important to realize that we are always connected at deeper levels, no matter what the surface of our lives may look like.

- Fear is a human egoic concept that interferes with our natural ability to experience the flow of eternal and ever-present Love. If we are fearful, we cannot be free. Fear is the opposite of Freedom.

- As long as we identify only with our physical form, we will embody fear. When we come to understand that we are not our bodies—that we are so much more than these bodies—then the fear spontaneously evaporates. Once we learn how to fully occupy our bodies without identifying with them, we can begin to live our lives to their maximum potential. Only then can our spirit fully express itself.

- Love always "is." Love is a sea within which we are always free to dwell; it is a place within—one of deep connection and one that we can always enter. Love has no opposite. We cannot give or share this type of love; instead, we simply open ourselves to experience the Love that is everywhere and always present.

- Freedom means no longer identifying with our body and ego because we recognize and deeply understand the temporary, illusionary nature of our form.

- The deepest expression of Love is for each of us to live our lives fully and express our own special uniqueness, as we also recognize and honor our deep, but direct, connection to all others.

234

CONCLUSION

INTRODUCTION

From within our three-dimensional reference frame, there can be no absolute answers to our most profound questions—those about the deeper nature of our lives. If there are such answers, they exist in realms or dimensions that lie beyond our *viewport* and our present ability to comprehend or explain. Our answers will continue to evolve, as we evolve.

Our personal life experience, our spiritual traditions, and our most recent science all share a profound common ground. Their intersection illuminates the existence of an amazing universe that lies far beyond any worlds that we understand or generally imagine. As we start to better understand our existence, we can compile a list of suggested recommendations that point towards a healthier and happier way of living on our planet: a way of being that will also change everything about how our lives appear. These techniques, and others like them, have already served many by helping to make their lives more purposeful, exciting, and joyful. What follows is a re-worded and re-ordered summary of key points from this book. Some of these points have not been fully developed in this abridged compilation, so if a reader seeks to understand any of these ideas in more depth, they should refer directly to *The Architecture of Freedom.*

DEATH, SELF, AND OUR VIEWPORT

If we were completely free to perceive the universe without the limitations of our *viewport,* we would directly experience much more of the *infinite* expanse of creation. Included in our expanded vista would be *infinite* possibilities for all futures and pasts, and every possible outcome of every thought or decision by every being. Such a dramatic *viewport* change would mean that we would perceive all things "at once," which is actually a perspective that can only be understood "outside" of time. We would also experience exactly how everything in creation fits together perfectly. We would also experience everything from everywhere, which is not a "place" that humans can understand. With our current abilities, if we suddenly became this aware of the fully interactive and multi-dimensional Web of Infinite Possibilities, our brains and sensory systems would instantly find themselves completely overwhelmed. Such an expanded view of "time and space" is far beyond our current capabilities.

From our current reference frame, it is impossible to imagine what this infinite and timeless *viewport* experience would be like. Because of our three-dimensional specialization, we completely lack the tools for understanding, utilizing or relating to this greater extent. Since we cannot directly see or experience our deep interconnection from our current *viewport* we, instead, often perceive chaos, randomness, accidents, and separation. However, certain aspects of this expanded experience are still describable, and understanding these can greatly enhance our experience.

If our personal viewport expanded, we would eventually reach a place where we would clearly understand that there are no separate individuals! We would discover that our persistent sense of each of us being individual and separate is only a grand illusion generated through our limited three-dimensional *viewport*. ***We would instantly understand that everything within our universe is directly connected to everything else and always responding through direct, intimate, and instantaneous cause-and-effect interactions. It would be clear that all things and occurrences in our universe are the result of this deep-level interconnectivity; there never are accidents and nothing is random or separate. Everything that unfolds is only the direct result of this invisible, yet intimate and deep, interconnection.***

From our new expanded *viewport,* we would also discover that even though we were not usually aware of it, there has always been some unchanging part of us fully aware of these deeper interconnections. This part is the *self*: that sacred place inside all of us where we hold the knowledge of our eternal connection to deeper levels of *being*.

With this expanded *viewport* we would understand how we are both one and many at the same time. This seems paradoxical, but this type of conceptual difficulty is only a byproduct of our dimensionally limited and logically constrained minds. The ancient philosophical question "do I exist?" completely depends on the meaning of "I." If our personal understanding of "I" implies complete separation and the continuation of our individual egos, then this question must be answered with a "no." At the deepest levels, the separate individual is recognized as just a temporary phenomenon: a brief occurrence within "time" that is fully confined to the three-dimensional reference frame. However, if the "I" is defined from that place where no separation exists, then not only does the "I" exist, but also it is infinite and eternal. From our *viewport* we have a very difficult time imagining that both of these manifestations can exist simultaneously. *We function as both an integrated part of a much bigger being and as an individual at the same time! Our collective experience of isolation and separation is only an illusion formed through our limited viewport.*

Since we are not separate individuals, there is no separate one who will die. Death, as it turns out, is also an illusion. Our thinking that death is inevitable may be the most confining illusion experienced within our viewport. Understanding the deep nature and significance of this illusion has profound implications.

IT ALL EXISTS FOR US

This book describes the amazing architecture of our universe, where everything that could possibly exist in what we think of as the "present, past and future," already exists. It all exists together, as a continuum, in the singular place and time called the "now." Included in this continuum is all that which is destructive and creative, terrible and wonderful—all of it—along with everything that ever was, ever could be, could have been and will be. It all exists at once, now and forever, in the infinite and multi-dimensional "Web of Infinite Possibilities."

This "Web of Infinite Possibilities" actually has some similarities to the modern Internet. In our computer-based universe, we simply type in a new address and instantly find ourselves shifted to a new place in the world wide web of the Internet. In the "Web of Infinite Possibilities," it is much the same, but changing locations requires something much deeper than just typing. Location can only be changed through the shifting of our inner *vibration*. While our thoughts may support these changes, our thoughts, by themselves, cannot change our location or the appearance of our world. A deeper type of inner communication is required.

Even if we knew how to access an "address" of some imagined place in this Web, we still would not be able to accurately predict the terrain or the type of adventure because of another important principle: *This Web is fully interactive and interconnected in a manner through which every new experience will become instantly unique, simply because we are now present and participating. All experience is shaped by our participation. Each of us is an integrated and critical part of our every encounter and interaction.*

This means that our personal experience is ours alone. In fact, there can be no experience without the participation of the experiencer. Because each person's participation is such an integral part of any experience, two people can never share the same experience; each of us will experience any event from our own unique perspective.

As we each learn how to free ourselves from the many layers of conditioning that bind us, we become more like a tourist who chooses where to travel. We learn how to more consciously explore, observe, and participate freely in this expanded playground. Through the opening and expansion of our inner *being,* we will become more aware of all the different aspects and parts of this beautiful, elegant, and unlimited creation. Every possibility will still always exist, but we will learn how to "choose" those experiences that have the most meaning for us in our present vibrational state. *This "choosing" is not executed by the mind but, is a result of the open flow from our inner being. The perfect choices will always arise from the uninhibited and natural resonant expression of our fundamental vibrational state of being at any given moment. For this to happen, we simply need to learn how to get out of our own way and relax to the natural flow of life.*

The process of learning how to better "navigate" in this landscape of many dimensions is not about becoming a "better" person. Understanding and using these ideas will not make us, or even our lives, better. *Each of us, and everything in the world, is already perfect, right now!* This critically important principle is also extremely ironic, paradoxical, and can be very difficult to internalize. *Understanding that everything is already perfect, exactly as it appears, is essential for the expansion of our freedom and self-empowerment.*

We are all engaged in this process of evolution and when the "time" is right, each of us will encounter just the perfect opportunities and challenges for the development of our unique aspect of *being*. This is the deep nature of this fully interactive process—all things unfold when the "time" is right. *The "right time" is when our individual inner vibration is perfectly tuned to interact or resonate with these particular experiences.*

As we evolve, all of the existing elements of our lives will still be present—life actually will not feel that different because it always involves expansion; a process that is inclusive, loses nothing and leaves nothing behind. As we evolve and expand, the context will change so we will become better able to see deeper connections and relationships. "Good" and "bad" will be seen for what they really are— integral elements of *duality* that are necessary parts of our physical creation.

At each and every step, we will still be presented with unexpected challenges; this is the nature of this evolutionary process. With a new awareness, however, these challenges can more easily be understood as nothing more than fresh opportunities to revisit, explore, and develop those aspects of our *being* that are critical for its full expression. *If we find ourselves in a situation that appears particularly difficult, we will understand that this is because some part of us is still resisting the natural flow of life, or asking for deeper growth.* Any internal resistance to what is presented in the moment will always create "difficulties," because resistance limits the full expression of our being. If we choose to, we can move beyond this resistance by witnessing, learning, knowing, and then, finally, incorporating these resisted parts into our full and loving embrace. This is a never-ending process, but it becomes much easier and more comfortable as we learn how not to resist, and how to flow more naturally with life's deeper rhythms.

THE LIMITS OF WHAT WE KNOW TODAY

Our current physical form is an expression that depends on "time," and functions using dualistic concepts. Our human physical form exists in a very small and well-contained sub-space within the Web of Infinite Possibilities. Some of these containing elements are obvious; we need air, water, food, gravity, and other environmental support systems. Some are less obvious; we require the structure of marching time and dualistic contrast to reason and function. This container is our womb, making it easy and natural to miss the broader perspective: one that would instantly reveal the larger picture about who and what we really are.

As physical beings, we will always be witnessing, experiencing, and describing life from this limited frame of reference. This is a fundamental truth about our experience, and the recognition of this truth is another foundation stone of this book. Our size, the *energy* expenditure of our processes, our limited senses, our conceptual abilities, and our physical location are all specific to this particular physical expression. It is impossible for us to fully understand what exists beyond the natural limits of our senses, abilities and awareness. We simply cannot expect to understand creation from this limited perspective.

We, therefore, have a natural built-in limit that we call our "conceptual horizon"—we simply are not able to conceive or fully understand anything that exists beyond this limit. To understand our universe in more depth, the extents of our *conceptual horizon* must first expand, which requires the full range of experience. This is life's evolutionary purpose.

With our current level of knowledge and technology, we do understand a few things well, and some of these things involve "what our universe is not." We have learned that it is not solid, it is not limited to three dimensions, and it is not limited by the one-directional "arrow of time." We, ourselves, are a part of this universe; therefore, we also are "not" these things. We are not three-dimensionally bound, individual, vulnerable, and very separate beings living in a hostile and competitive world. While we may not be able to say with any certainty exactly who or what we are, we can say with the certainty of our physics, spiritual traditions, and life experience that we are not limited by time or space in the ways that we once thought.

THE PERSISTANCE OF THE ILLUSION

The most common explanation for the non-*probabilistic* (or *deterministic*) appearance of our human-scaled physical world is that even though each particle is *probabilistic* and unpredictable, because so many are involved in every interaction, they average out to produce the very consistent and predictable results that we observe. A similar thing can be said about our beliefs. The more often a belief is communicated through vibration, the more likely it is to be seen as the projected "reality" of our realm. As individuals, we have much more control of our beliefs so change can happen in an instant. With a larger group, "averaging" occurs, making cultural change a slow, gradual, and sometimes very difficult process. Collectively, we all contribute to the "averaging" that results in the persistence of the illusion that we experience as our reality. However, each individual is fully capable of breaking free from this normalization to enter a new world where a different "averaging" occurs.

HERE FOR THE EXPERIENCE

An important living principle derived from this architecture is that we all are, in each and every moment, contributing important and integrated functions within this amazing evolutionary process called life. ***This means that we are always fulfilling an essential role, no matter how our lives may appear.***

Nothing about our lives ever needs to be changed because we are always growing and evolving through every little thing we do and experience. Personal and collective shifts in evolution happen in a natural and orderly process when the time and circumstances are right. Life is always showing us how to become more compassionate, and this is a continuous influence on our evolution. Everything that we meet with in our lives adds to the richness of our adventure and contributes to our evolutionary progress. ***One function of our physical world is to provide us with the raw material for the shaping of our soul in preparation for our next stage of evolution. Immersed in this process, we are always living an important lifetime, and we are always contributing to the whole of creation, no matter how our particular situation may appear. We each are presented with the opportunity to experience only a small subset of this amazing creation, but it always contains the perfect encounters, relationships, problems, and information that we need for our personal evolution.***

The challenge for each individual, for all of us, is to be authentic and true to our self. That purpose ultimately invites us to live our lives here on Earth as consciously and with as much gratitude and love as is possible—this is the most "efficient" way to participate fully, and involves the least amount of suffering. We are here to make acquaintance with all that this life has to offer—all of the

extremes offered by this dualistic realm—and then to use our experiences to advance and broaden our collective evolution.

We can do no better than to live a grateful and authentic life. Such an existence allows us to become fully acquainted with love, anger, the beautiful, the not-so-beautiful, happiness, sadness, joy, despair, ecstasy, and emptiness. Every personality, every sensation, every misstep, every correction, every part of life is equally important to live and witness. There are no mistakes or wrong turns because through this deep and personal participation in life, we gradually become, both individually and collectively, blessed with a much greater wisdom: the wisdom of direct and personal experience.

Guided by this experiential wisdom, we gradually make the internal changes that affect our individual and collective evolution. We can only make these changes when we are fully prepared; our honest, conscious, and heartfelt participation, in all aspects of life, provides this preparation.

Understanding that everything is exactly right in each and every moment is very difficult for the human mind. The empowering paradox is that, at the same time, we all have the hidden ability to shape our adventure in an infinite number of dramatic ways.

THE NEW TERRAIN

While the *cosmological* possibilities dramatically increase with the addition of extra dimensions, a universe of ten or eleven dimensions is not a space that we can ever "understand" from our *viewport*. Even the word "space," has powerful three-dimensional associations that try to pull us back into our conceptual mindset., We can experience new forms of "space" by approaching our "thinking" differently, but only if these are close enough to our *viewport* to be still within our ever-expanding *conceptual horizon*. Living within these new bounds requires a much more inter-connected and flowing mindset because this new expanded space not only contains much more *information,* but also its inter-connections will not appear to be rational. To access this different way of being, we need to first understand, and then relax, our controlling cultural conditioning and habitual thinking. We all are capable of sensing more aspects of this invisible world, but we can only develop the ability to do so through direct experience and the expansion of our awareness.

We all have occasional "events" in our lives, sometimes traumatic, which, unexpectedly, open temporary windows to this unseen world. These "out-of-the-ordinary" occurrences may seem random or accidental but they are not, because of the interactive and fully responsive nature of the *multiverse*. However, brief as they may be, these types of encounters can imprint and inform us for the rest of our lives.

Even though we may not see or easily understand the vast hidden depths of our universe, we are still able to benefit directly from its amazing architecture. If all that we each do is accept its *infinite* extents and be open and willing to explore the possibilities that are presented, then we will benefit in extraordinary ways because this human playground, while enormous and mysterious, is always intimate, and fully connected to us beyond *time* and *space*. Even though our personal *viewports* are limited, we are always functioning as an integrated and critical part of everything, continuously receiving *information* from beyond the shadows of our direct experience. *Information* fills this amazing and expansive space, and with the right kind of temperament and practice, we become available to more of it.

As we open our hearts to these deeper types of perception and knowledge, we have no certainty about what this new *information* and our open participation will really mean. If we are to "understand" our experiences at this deeper level, we must first allow for the relaxation of our normal conceptual thinking.

The oft-quoted quantum physicist, Niels Bohr, was famous for criticizing new *quantum* sub-theories and ideas because they were, in his words, *"not crazy enough."* We have learned through our history with other major cultural shifts that at first, the new ideas often appear to be "crazy" but then, as we evolve, we begin to see them through a new and different lens—one in which all the pieces fit together better and those "crazy ideas" suddenly make sense! We also understand through this same history that it will take "time" for our culture to assimilate this new knowledge and undergo a larger-scale awakening. If history is an accurate indicator, the current paradigm shift could take several hundred years to complete. (This was the amount of time required for the last major paradigm shift, the one that moved humankind from the "flat earth," to our current *heliocentric viewport.)*

However, it is important to understand that this entire discussion about a more gradual evolution assumes that we interact with "time" in our old way. If we also allow our perception of "time" to shift, then this radical change can occur in an instant.

When the process of scientific discovery is viewed from within our old paradigm of linear time, it seems that our science is simply uncovering previously unknown things about our world. There is, as discussed, another way to view this process of discovery. Another possible view, one that understands the *Many Worlds theory,* is that **we may be "creating" this physics and these newly discovered parts of our universe on the fly, as we collapse the possibilities and probabilities into a new physical form with our thoughts and experimental observations.** From our normal cultural *viewport,* a creative process like this seems rather fantastic because it means that our actual physical universe will always be evolving and expanding as rapidly as our collective vibrations can "create" it. We actually "dream" our universe into a physical reality. Of course, this also means that, from the perspective of linear time, it will always be impossible to fully describe our physical universe because it will always be a moving target that is continuously evolving to include whatever we can "discover" or imagine next.

However, if we view this same process of "discovery equals creation" from an even broader perspective, that of the timeless Multiverse, then all of these possibilities already and always exist within the fabric of creation. The process is now understood to be only our opening to and recognition of something that always was present, but hidden beyond our viewport. As we discover what it means to view this unfolding process from the timeless perspective of the Multiverse, many things that we once considered strange or mysterious, suddenly begin to make sense.

THE CAVEAT

This entire book is written under the shadow of one enormous caveat: *No matter what we explore, imagine, or dream, all the principles of physics are ultimately only ideas in our minds. Concepts do not exist outside of the conceptualizers.*

As Albert Einstein wrote: "Physical concepts are free creations of the human mind, and are not, however it may seem, uniquely determined by the external world."[4]

It is impossible for us to "think" far outside the box of our conceptual minds, especially since the entire universe as we understand it is only a mental concept of our own creation. Today, from our Earth-bound and limited perspective, it appears that as we learn more about our world, the physics only gets more complicated! However, this complexity in our mathematical descriptions of the universe is really only a symptom of an immature theory. As we learn more and begin to witness creation from a broader perspective, our understanding will become more inclusive, and things will become simpler, not more complicated. Our present confusion and complications are only phenomena being shaped by the limits of our *viewport*. As our view expands, a more elegant, simple, beautiful, and unified vision of our existence will be gradually revealed.

Michio Kaku is one of our most brilliant contemporary physicists and an outstanding communicator. For many years he has been demonstrating his uncanny ability to "explain the unexplainable" with statements such as, "Although the mathematical complexity of the ten-dimensional theory has soared to dizzying heights, opening up new areas of mathematics in the process, the basic ideas driving unification forward, such as higher dimensional space and strings, are fundamentally simple and geometric." To him and many in the physics community, the present-day appearance of complexity is a sure indicator that we are still working with incomplete and immature theories.

As we explore the physics of ten, eleven, or more dimensions, one thing that will continue to be problematic is obtaining scientific proof for these theories. Human beings completely lack the conceptual framework and tools to perform experiments in these types of spaces. Exploring physics, life, or consciousness within these unknown regions requires a kind of undaunted faith that is reminiscent of the original new-world explorers. Lacking charts and directions, they had to trust a personal vision that often tapped deeply into their religious, spiritual and personal beliefs.

---

[4] *The Evolution of Physics* (1938), co-written with Leopold Infeld.

The place where science, spirituality, and experience intersect is an unknown territory brimming with potential riches from life's deepest secrets. *Any scientific journey seeking truth will necessarily converge with honest spiritual explorations and our actual life experience, since all three point towards the same truth: the deepest secret of life.*

CLOSING

As we explore, evolve, and learn more about ourselves and our special relationship to the universe, we are also discovering that the mysterious and unknown parts of creation also appear to expand. These unknown regions have already become, by far, the greatest and most dominant component of our new scientific awareness. The mystery seems to grow exponentially each and every time we include a new insight that expands our previous vision. The more we know, the less it seems we really understand.

The *Multiverse* can never be understood or explained using the current language of our minds. Creation is far too *infinite* to be contained by our limited concepts or words. Any "thinking" will only lead to misconceptions, so this book, which must be built upon thoughts and words, may even encourage such misconceptions. Our "ideas" about this amazing expanse will never be accurate, and these new ideas can even become actual limiting concepts that bind us to this realm and further restrict our freedom. At best, our words and concepts can be temporary pointers that help guide us towards a deeper, but still relative, truth.

Our conceptual brains always build upon that which we already understand; this is how we function. Because the Multiverse extends so far beyond our conceptual space and understanding, any direct or logical comparison to known things and ideas will always fall short and limit the possibilities. However, just the awareness that we are not capable of fully comprehending the depths of the *Multiverse* is, by itself, a significant realization—one that can help us begin to let go of our confining need to "understand." It is largely from remaining open to these "other possibilities" that we can first glimpse and then learn about the deeper truths of creation.

As we continue to let go of our old concepts and open ourselves to our *infinite* possibilities, we will uncover new ways to explore, occupy and enjoy our expansive *Multiverse*. **This amazing, infinite and evolutionary journey of Being has always been, and will always be, our collective destiny and the ultimate purpose of life.**

RESOURCES

(Please feel free to recommend other resources.)

PHYSICS, MATH AND QUANTUM PHYSICS

Albert Einstein
Essays in Physics
1950, Philosophical Library

Out of My Later Years
1956, Citadel Press

Joseph Schwartz and Michael McGuiness
Einstein for Beginners
1979, Pantheon Books

Francis S. Collins, head of Human Genome Project
The Language of God
2006, Free Press

Brian Greene
The Elegant Universe: Superstrings, Hidden Dimension, and the Quest for the Ultimate Theory

The Fabric of the Cosmos: Space, Time and the Texture of Reality
2004, Knopf

The Hidden Reality: Parallel Universes and the Deep Laws of the Cosmos
2011, Alfred A. Knopf

Michio Kaku
Beyond Einstein
1997, Oxford University Press

Visions
1999, Oxford University Press

Hyperspace

1994, Oxford University Press

Timothy Ferris
The Whole Shebang: A State of the Universe(s) Report
1997, Simon and Schuster

Bertrand Russell
The ABC of Relativity
1958, George Allen

Tom Siegfried
Strange Matters
2002, Joseph Henry Press

Lawrence Krauss
Lecture "A Universe from Nothing"
You-Tube

ASTRONOMY AND COSMOLOGY

William J. Kaufmann, III
Discovering the Universe
1987, W.H Freeman and Company

Carl Sagan
Cosmos
1980, Random House

ANTHROPOLOGY

Robert Ardrey
African Genesis
1961, Macmillan

Territorial Imperative
1966, Kodansha Globe

META PHYSICS

Fritjof Capra
The Tao of Physics
1975, Bantam Books

The Web of Life
1996, Anchor Books-Doubleday

The Turning Point
1982, Bantam Books
Movie titled "Mindwalk" directed by Bernt Capra

David Darling
Equations of Eternity
1993, Hyperion

Zen Physics–The Science of Death, The Logic of Reincarnation
1996, Harper Collins

Frank J Tipler
The Physics of Immortality
1994, Doubleday

Frank Wilczek with Betsy Devine
Longing For the Harmonies
1987 W.W. Norton Co.

Fred Alan Wolf Ph.D.
Many books, including:

The Dreaming Universe
1994, Simon and Schuster

Eagle's Quest
1991, Touchstone-Simon and Schuster

Parallel Universes
1988, Touchstone

Paul Davies
Many books, including:

Are We Alone
1995, Orion Publications

About Time
1995, Orion Publications

The Mind of God
1992, Orion Productions

Space and Time in the Modern Universe
1977, Cambridge University Press

PHILOSOPHY OF SCIENCE

Gary Zukav
The Seat of the Soul
1989, Fireside

The Dancing Wu Li Masters
1979, Bantam

Robert M Pirsig
Zen and the Art of Motorcycle Maintenance
1974, Bantam

Ken Wilber
A Brief History of Everything
1996, Shambahla

The Holographic Paradigm
1982, New Science Library

Eugene Pascal Ph.L.
Jung to Live By
1992, Warner Books

Krista Tippett-Interviews
Einstein's God-Conversations about Science and the Human Spirit
2010, Penguin

Steve McIntosh
Integral Consciousness
2007, Paragon House

PSYCHOLOGY

Stanislav Grof M.D.
The Holotropic Mind: Three Levels of Human Consciousness
1993, Harper-San Francisco

Brian L. Weiss M.D.
Same Soul, Many Bodies
2004, Free Press

Many Lives, Many Masters
1988, Simon and Schuster

Only Love Is Real
1997, Grand Central Publishing

RESEARCH SCIENCE

**Rollin McCraty, Ph.D., Executive Vice President and Director
The Institute of HeartMath**
Numerous research papers
Quoted in Movie "I Am"

Lewis Thomas
The Lives of a Cell
1974, Viking Press

J. Konrad Stettbascher
Making Sense of Suffering
1993, Meridian

MUSIC

Robert Jourdain
Music, the Brain and Ecstasy
1997, Bard Press

Daniel J. Levitin
This Is Your Brain on Music
2006, Plume

FICTION

Edwin A Abbott
Flatland
1952, Dover Publications

Michael Murphy
Golf in the Kingdom
199,7 Arkana

Jacob Atabet
1977, Jeremy P. Thacher

NON-FICTION

Carol Riddell
The Findhorn Community
1990, Findhorn Press

Elizabeth Gilbert
Eat, Pray, Love
2007, Penguin Books

POETRY AND ART

Kahlil Gibran
Between Morning and Light
1972, Philosophical Library

The Prophet
1923, Alfred A Knopf

SPIRITUALITY AND RELIGION

Western

Elaine Pagels
The Gnostic Gospels
1979, Vintage Books

Jean–Yves Leloup
The Gospel of Philip
2003, Inner Traditions

Michael Wise, Martin Abegg Jr. and Edward Cook, editors and translators
Dead Sea Scrolls
1996, Harper: San Francisco

James M. Robinson, editor
The Nag Hammadi Library
1978, Harper and Row

Stephan A Hoeller
Jung and the Lost Gospels: Insights into the Dead Sea Scrolls
1989, The Theosophical Publishing House

Kyriacos C. Markides
Riding with the Lion: The Search of Mystical Christianity
1995, Penguin

Mathew Fox
One River, Many Wells
2000, Tarcher/Putnam

The Coming of the Cosmic Christ
1988, Harper Collins

Rupert Sheldrake with Matthew Fox
The Physics of Angels: Where Science and Spirit Meet
1996, Harper: San Francisco

Eckhart Tolle
The Power of Now
1999, New World Library

The New Earth: Awakening to Your Life's Purpose
2005, Dutton Publishing

Anthony deMello
Awareness
1990, Doubleday

The Way of Love: The Last Meditations of Anthony deMello
1991, Doubleday

William Dych. S.J., editor
Anthony DeMello writings
1999, Orbis Books

Edward Conze, editor and translator
Buddhist Scriptures
1959, Penguin Books

Jack Kornfield
After the Ecstasy, the Laundry: How the Heart Grows Wise on the Spiritual Path
2000, Bantum

Stephen Mitchell, editor
The Enlightened Heart, sacred poetry
1989, Harper and Row

Paul Ferrini
Reflections of the Christ Mind
2000, Doubleday

Nicole Gausseron
The Little Notebook: The Journal of a Contemporary Woman's Encounters with Jesus
1995, Harper San Francisco

Joan Borysenko Ph.D.
The Ways of the Mystic: 7 Paths to God
1997, Hay House

Sam Keen
To a Dancing God
1970, Harper

Helen Schucman and William Thetford
A Course in Miracles
1975, Foundation for Inner Peace

Eastern

Thich Nhat Hanh
Peace Is in Every Step: The Path of Mindfulness in Everyday Life
1991, Bantam

Swami Prabhavananda and Christopher Isherwood, editors and translators
How to Know God: the Yoga Aphorisms of Patanjali
1953 & 1981,Vendanta Press

Christopher Isherwood
Vendanta for the Western World
1945, Vendata Society of Southern California

Paramahansa Yogananda
The Autobiography of a Yogi
1946, Self-Realization Fellowship

J. Krishnamurti
Krishnamurti's Journal
1982, Harper and Row

Think on These Things
1981, Harper One

You Are the World
2001, Krishnamurti Foundation

Freedom from the Known
1975, Harper Colins

Stuart Holroyd
Krishnamurti: The Man the Mystery & the Message
1991, Element

Lao Tzu
Tao Te Ching
1997, Wordsworth Editions

Birgitte Rodriguez
Glimpses of the Divine: Working with the Teachings of Sai Baba
1993, Samuel Weiser

Howard Murphet
Sai Baba Man of Miracles
1971, Samuel Weiser Inc.

Ramana Maharshi
The Spiritual Teaching of Ramana Maharshi
1972, Shambala Publications

Adyashanti
The Way of Liberation
2012, Open Gate Sangha Inc.

Kahil Gibran
The Prophet
1979, Alfred A. Knopf

The Voice of the Master
1958, Bantam

Satyam Nadeen
From Onions to Pearls: A Journal of Awakening and Deliverance
1996, Hay House

From Seekers to Finders
2000, Hay House

Gopi Krishna
Living With Kundalini: The Autobiography of Gopi Krishna
1993, Shambahla

Paul Lowe
In Each Moment: A New Way to Live
1998, Looking-Glass Press

The Experiment Is Over
1989, New York

Shri Nisargadatta Maharaj
I Am That
1982, Acorn Press

Ram Dass
Be Here Now
1971, Lama Foundation

Still Here
2000, Riverhead Books

The Path of Service, audio book
1990, Sounds True Recordings

Bubba Free John
The Knee of Listening
1972, Dawn Horse Press

Nirmala-Daniel Erway
Nothing Personal: Seeing Beyond the Illusion of a Separate Self
2001, Endless Satsang Press

Sri H.W.L. Poonja-Papaji
THIS: Prose and Poetry of Dancing Emptiness
2000, Samuel Weiser

Eli Jaxon-Bear, editor
Wake Up and Roar: Satsang with H.W.L. Poonja
1992, Gangaji Foundation

No index entries found.